Multiplicative
Differential Calculus

Multiplicative
Differential Calculus

Svetlin G. Georgiev
Khaled Zennir

CRC Press
Taylor & Francis Group
Boca Raton London New York

CRC Press is an imprint of the
Taylor & Francis Group, an **informa** business
A CHAPMAN & HALL BOOK

First edition published 2023
by CRC Press
6000 Broken Sound Parkway NW, Suite 300, Boca Raton, FL 33487-2742

and by CRC Press
4 Park Square, Milton Park, Abingdon, Oxon, OX14 4RN

CRC Press is an imprint of Taylor & Francis Group, LLC

ISBN: 978-1-032-28912-0 (hbk)
ISBN: 978-1-032-28913-7 (pbk)
ISBN: 978-1-003-29908-0 (ebk)

DOI: 10.1201/9781003299080

Typeset in font CMR10
by KnowledgeWorks Global Ltd.

Contents

Preface

Differential and integral calculus, the most applicable mathematical theory, was created independently by Isaac Newton and Gottfried Wilhelm Leibnitz in the second half of the 17th century. Later, Leonard Euler redirected calculus by giving a central place to the concept of function, and thus founded analysis. Two operations, differentiation and integration, are basic in calculus and analysis. In fact, they are the infinitesimal versions of the subtraction and addition operations on numbers, respectively. In the period from 1967 till 1970, Michael Grossman and Robert Katz gave definitions of a new kind of derivative and integral, moving the roles of subtraction and addition to division and multiplication, and thus established a new calculus, called multiplicative calculus. Sometimes, it is called an alternative or non-Newtonian calculus as well. Multiplicative calculus can especially be useful as a mathematical tool for economics and finance.

This book is devoted to the multiplicative differential calculus. It summarizes the most recent contributions in this area. The book is intended for senior undergraduate students and beginning graduate students of engineering and science courses. The book contains seven chapters. The chapters in the book are pedagogically organized. Each chapter concludes with a section of practical problems. In Chapter 1, we introduce the field $R_\star$ and define the basic multiplicative arithmetic operations: multiplicative addition, multiplicative subtraction, multiplicative multiplication and multiplicative division, and are given some of their properties. In this chapter, we have defined the basic elementary multiplicative functions and are deducted some of their properties. Chapter 2 is devoted on the multiplicative derivative of a function. We deduct some of its properties such as multiplicative differentiation of multiplicative sum of two functions, multiplicative differentiation of multiplicative product and multiplicative quotient of two functions. In this chapter, we have defined multiplicative differentials and given some criteria for monotonicity of a function and local extremum of a function. In this chapter, we have deducted the multiplicative Rolle theorem, Lagrange theorem and Cauchy theorem as well as the multiplicative Taylor formula. In Chapter 3, we introduce the indefinite multiplicative integral and the Cauchy multiplicative integral and deduct some of their properties. We deduct the table of the basic multiplicative integrals. We consider multiplicative integration by substitutions and multiplicative integration by parts. We prove some important inequalities for the multiplicative integrals and deduct and prove some mean value theorems for multiplicative integrals. Chapter 4 deals with the improper multiplicative integrals on finite and infinite intervals. We give and prove some criteria for convergence and divergence of improper multiplicative integrals. In Chapter 5, we introduce the space $\mathbb{R}^n_\star$ and define in the basic multiplicative operations:

multiplicative addition and multiplicative multiplication, and proved that $\mathbb{R}_\star^n$ is a linear vector space. We define the multiplicative inner product of multiplicative vectors, multiplicative length of a multiplicative vector and multiplicative distance between two multiplicative vectors and deduct some of their properties. In Chapter 6, we define partial multiplicative derivatives of first and higher order and investigate some of their properties. In this chapter, we have considered multiplicative differentials and are given their expressions. We define multiplicative directional derivatives and deduct some of their properties. In this chapter, we have given some necessary conditions for existence of local extremum of a function. In Chapter 7, we investigate multiplicative integrals depending on parameters and using them and define iterated multiplicative integrals. We define multiple multiplicative and multiple improper multiplicative integrals and investigate their properties.

This book is addressed to a wide audience of specialists such as mathematicians, physicists, engineers and biologists. It can be used as a textbook at the graduate level and as a reference book for several disciplines.

Svetlin G. Georgiev and Khaled Zennir
Paris

Authors

Svetlin G. Georgiev is a mathematician who has worked in various areas of the study. He currently focuses on harmonic analysis, functional analysis, partial differential equations, ordinary differential equations, Clifford and quaternion analysis, integral equations, and dynamic calculus on time scales. He is also the author of *Dynamic Geometry of Time Scales*, CRC Press/Taylor & Francis Group. He is a coauthor of *Conformable Dynamic Equations on Time Scales*, with Douglas R. Anderson, CRC Press/Taylor & Francis Group.

Khaled Zennir earned his PhD in mathematics from Sidi Bel Abbès University, Algeria. He received his highest diploma in Habilitation in mathematics from Constantine University, Algeria. He is currently Assistant Professor at Qassim University in the Kingdom of Saudi Arabia. His research interests lie in the subjects of nonlinear hyperbolic partial differential equations: global existence, blowup, and long-time behavior.

The authors have also published: *Multiple Fixed-Point Theorems and Applications in the Theory of ODEs, FDEs and PDE; Boundary Value Problems on Time Scales, Volume 1* and *Volume II*, all with CRC Press/Taylor & Francis Group.

1

The Field $\mathbb{R}_\star$

In this chapter, we introduce the field $\mathbb{R}_\star$ and define the basic multiplicative arithmetic operations: multiplicative addition, multiplicative subtraction, multiplicative multiplication and multiplicative division and are given some of their properties. In this chapter, we define the basic elementary multiplicative functions and are also deducted some of their properties.

1.1 Definition

Let $\mathbb{R}_\star = (0, \infty)$.

Definition 1.1. In the set $\mathbb{R}_\star$ we define the multiplicative addition or $\star$ addition $+_\star$ in the following manner
$$a +_\star b = ab, \quad a, b \in \mathbb{R}_\star.$$

Example 1.1. Let $a = 1$, $b = 3$. Then
$$a +_\star b = 1 +_\star 3$$
$$= 1 \cdot 3$$
$$= 3.$$

Definition 1.2. In the set $\mathbb{R}_\star$ we define the multiplicative multiplication or $\star$ multiplication $\cdot_\star$ as follows
$$a \cdot_\star b = e^{\log a \log b}.$$

Example 1.2. Let $a = 1$ and $b = e$. Then
$$a \cdot_\star b = e^{\log 1 \log e}$$
$$= 1.$$

DOI: 10.1201/9781003299080-1

Example 1.3. Let $a = 2$, $b = \dfrac{1}{3}$, $c = 4$. We will find

$$A = (a +_\star b) \cdot_\star c.$$

We have

$$
\begin{aligned}
a +_\star b &= 2 \cdot \frac{1}{3} \\[2mm]
&= \frac{2}{3}.
\end{aligned}
$$

Then

$$
\begin{aligned}
A &= e^{\log(a +_\star b)\log c} \\[2mm]
&= e^{\log \frac{2}{3} \log 4} \\[2mm]
&= e^{2\log 2 \log \frac{2}{3}}.
\end{aligned}
$$

Exercise 1.1. Let $a = e^3$, $b = e^4$, $c = e^{10}$. Find

$$A = (a +_\star b) \cdot_\star c.$$

Answer 1.1. $A = e^{70}$.

Definition 1.3. In the set $\mathbb{R}_\star$ we define the multiplicative zero ($\star$ zero) and multiplicative unit ($\star$ unit) as follows

$$0_\star = 1 \quad \text{and} \quad 1_\star = e.$$

Below, we have listed some of the properties of the multiplicative addition and multiplicative multiplication.

1. **Commutativity of $\star$ Addition.** Let $x, y \in \mathbb{R}_\star$ be arbitrarily chosen. Then

$$
\begin{aligned}
x +_\star y &= xy \\[2mm]
&= yx \\[2mm]
&= y +_\star x.
\end{aligned}
$$

2. **Associativity of $\star$ Addition.** Let $x, y, z \in \mathbb{R}_\star$ be arbitrarily chosen. Then

$$x +_\star (y +_\star z) = x +_\star (yz)$$

$$= xyz$$

$$= (xy)z$$

$$= (x +_\star y)z$$

$$= (x +_\star y) +_\star z.$$

3. $\star$ **Identity Element of** $\star$ **Addition.** Let $x \in \mathbb{R}_\star$ be arbitrarily chosen. Then

$$x +_\star 0_\star = x +_\star 1$$

$$= x \cdot 1$$

$$= x.$$

4. $\star$ **Inverse Elements of** $\star$ **Addition.** Let $x \in \mathbb{R}_\star$ be arbitrarily chosen. Define

$$-_\star x = \frac{1}{x}.$$

Then

$$x +_\star (-_\star x) = x +_\star \left(\frac{1}{x}\right)$$

$$= x \cdot \frac{1}{x}$$

$$= 1$$

$$= 0_\star.$$

5. $\star$ **Identity Element of** $\star$ **Multiplication.** Let $x \in \mathbb{R}_\star$ be arbitrarily chosen. Then

$$x \cdot_\star 1_\star = x \cdot_\star e$$

$$= e^{\log x \log e}$$

$$= e^{\log x}$$

$$= x.$$

6. ⋆ **Inverse Elements of** ⋆ **Multiplication.** Let $x \in \mathbb{R}_\star$ be arbitrarily chosen. Take
$$x^{-1\star} = e^{\frac{1}{\log x}}.$$

Then

$$
\begin{aligned}
x \cdot_\star x^{-1\star} &= x \cdot_\star \left(e^{\frac{1}{\log x}} \right) \\[2mm]
&= e^{\log x \log e^{\frac{1}{\log x}}} \\[2mm]
&= e^{\frac{\log x}{\log x}} \\[2mm]
&= e \\[2mm]
&= 1_\star.
\end{aligned}
$$

7. **Distributivity.** Let $x, y, z \in \mathbb{R}_\star$ be arbitrarily chosen. Then

$$
\begin{aligned}
(x +_\star y) \cdot_\star z &= (xy) \cdot_\star z \\[2mm]
&= e^{\log(xy) \log z} \\[2mm]
&= e^{(\log x + \log y) \log z} \\[2mm]
&= e^{\log x \log z + \log y \log z} \\[2mm]
&= e^{\log x \log z} e^{\log y \log z} \\[2mm]
&= (x \cdot_\star y) \cdot (y \cdot_\star z) \\[2mm]
&= (x \cdot_\star z) +_\star (y \cdot_\star z).
\end{aligned}
$$

Definition 1.4. For any $x \in \mathbb{R}_\star$, the number

$$-_\star x = \frac{1}{x}$$

will be called the multiplicative opposite number or ⋆ opposite number of x.

We have

$$-_\star(-_\star x) = -_\star \left(\frac{1}{x} \right)$$

$$= \frac{1}{\frac{1}{x}}$$

$$= x$$

for any $x \in \mathbb{R}_\star$.

Definition 1.5. For $x, y \in \mathbb{R}_\star$, define multiplicative subtraction or $\star$ subtraction $-_\star$ as follows

$$x -_\star y \;=\; x +_\star (-_\star y)$$

$$=\; x(-_\star y)$$

$$=\; x \cdot \frac{1}{y}$$

$$=\; \frac{x}{y}.$$

Definition 1.6. For $x \in \mathbb{R}_\star$, $x \neq 0_\star$, the number

$$x^{-1_\star} = e^{\frac{1}{\log x}}$$

will be called the multiplicative reciprocal or $\star$ reciprocal of the number x.

We have

$$\left(x^{-1_\star}\right)^{-1_\star} \;=\; \left(e^{\frac{1}{\log x}}\right)^{-1_\star}$$

$$=\; e^{\log e^{\frac{1}{\frac{1}{\log x}}}}$$

$$=\; e^{\log x}$$

$$=\; x$$

for any $x \in \mathbb{R}_\star$, $x \neq 0_\star$.

Definition 1.7. For $x, y \in \mathbb{R}_\star$, define multiplicative division or $\star$ division $/_\star$ as follows

$$x /_\star y \;=\; x \cdot_\star \left(y^{-1_\star}\right)$$

$$= x \cdot_\star \left(e^{\frac{1}{\log y}} \right)$$

$$= e^{\log x \log e^{\frac{1}{\log y}}}$$

$$= e^{\frac{\log x}{\log y}}.$$

Example 1.4. We will find

$$A = (2 +_\star 3) \cdot_\star 4 -_\star (3 +_\star 1)/_\star 5.$$

We have

$$
\begin{aligned}
A &= (2 \cdot 3) \cdot_\star 4 -_\star (3 \cdot 1)/_\star 5 \\
&= 6 \cdot_\star 4 -_\star 3/_\star 5 \\
&= e^{\log 6 \log 4} -_\star e^{\frac{\log 3}{\log 5}} \\
&= e^{2 \log 6 \log 2 - \frac{\log 3}{\log 5}}.
\end{aligned}
$$

Exercise 1.2. Find

1. $A = (3 -_\star 5)/_\star 2 +_\star (4 +_\star 2) \cdot_\star e.$
2. $A = 3 +_\star 2 -_\star 3 \cdot_\star (2 +_\star 4).$
3. $A = 1 -_\star 3 +_\star 4 \cdot_\star (1 +_\star 5).$

Answer 1.2.

1. $e^{\frac{\log \frac{3}{5}}{\log 2} + 3 \log 2}.$
2. $e^{\frac{\log 6}{3 \log 3 \log 2}}.$
3. $\frac{1}{3} e^{2 \log 2 \log 5}.$

Theorem 1.1. *For any* $a, b \in \mathbb{R}_\star$, *the equation*

$$a +_\star x = b \tag{1.1}$$

has at least one solution.

Proof Let

$$x = b -_\star a.$$

Then

$$a +_\star (b -_\star a) = a +_\star \left(\frac{b}{a} \right)$$

$$= a \cdot \frac{b}{a}$$

$$= b.$$

This completes the proof.

Corollary 1.1. Any solution $x \in \mathbb{R}_\star$ of the equation

$$a +_\star x = a, \quad a \in \mathbb{R}_\star, \tag{1.2}$$

is a solution of the equation

$$b +_\star x = b, \quad b \in \mathbb{R}_\star. \tag{1.3}$$

Proof By Theorem 1.1, it follows that the equation (1.2) and the equation

$$a +_\star y = b$$

have at least one solution x and y, respectively. Then

$$
\begin{aligned}
b +_\star x &= (a +_\star y) +_\star x \\[2mm]
&= a +_\star (y +_\star x) \\[2mm]
&= a +_\star (x +_\star y) \\[2mm]
&= (a +_\star x) +_\star y \\[2mm]
&= a +_\star y \\[2mm]
&= b,
\end{aligned}
$$

i.e., x is a solution of the equation (1.3). This completes the proof.

Corollary 1.2. The equation (1.2) has a unique solution.

Proof By Theorem 1.1, it follows that the equation (1.2) has at least one solution. Assume that the equation (1.2) has two solutions x and y. Then

$$a +_\star x = a$$

and

$$a +_\star y = a.$$

By Corollary 1.1, it follows

$$y +_\star x = y$$

and

$$x +_\star y = x.$$

Hence,

$$
\begin{aligned}
y &= y +_\star x \\
&= x +_\star y \\
&= x.
\end{aligned}
$$

This completes the proof.

Remark 1.1. *By Corollary 1.2, it follows that the multiplicative zero $0_\star$ is unique.*

Corollary 1.3. For any $a, b \in \mathbb{R}_\star$, the equation (1.1) has a unique solution.

   Proof By Theorem 1.1, it follows that the equation (1.1) has at least one solution. Assume that the equation (1.1) has two solutions x and y. Then

$$a +_\star x = b$$

and

$$a +_\star y = b.$$

By Theorem 1.1, it follows that the equation

$$a +_\star z = 0_\star$$

has at least one solution. Hence,

$$
\begin{aligned}
y &= y +_\star 0_\star \\
&= y +_\star (a +_\star z) \\
&= (y +_\star a) +_\star z \\
&= (a +_\star y) +_\star z \\
&= b +_\star z \\
&= (a +_\star x) +_\star z
\end{aligned}
$$

$$= a +_\star (x +_\star z)$$

$$= a +_\star (z +_\star x)$$

$$= (a +_\star z) +_\star x$$

$$= 0_\star +_\star x$$

$$= x.$$

This completes the proof.

Theorem 1.2. *For any* $a \in \mathbb{R}_\star$, $a \neq 0_\star$, *the equation*

$$a \cdot_\star x = b \tag{1.4}$$

has a solution.

   Proof Let

$$x = b /_\star a.$$

Then

$$
\begin{aligned}
a \cdot_\star x &= a \cdot_\star (b /_\star a) \\[2ex]
&= a \cdot_\star e^{\frac{\log b}{\log a}} \\[2ex]
&= e^{\log a \log e^{\frac{\log b}{\log a}}} \\[2ex]
&= e^{\log a \cdot \frac{\log b}{\log a}} \\[2ex]
&= e^{\log b} \\[2ex]
&= b.
\end{aligned}
$$

This completes the proof.

Corollary 1.4. If $a \in \mathbb{R}_\star$, $a \neq 0_\star$, then any solution of the equation

$$a \cdot_\star x = a \tag{1.5}$$

is a solution to the equation

$$b \cdot_\star x = b, \quad b \in \mathbb{R}_\star. \tag{1.6}$$

Proof By Theorem 1.2, it follows that the equation (1.5) and the equation

$$a \cdot_\star y = b$$

have at least one solution x and y, respectively. Then

$$
\begin{aligned}
b \cdot_\star x &= (a \cdot_\star y) \cdot_\star x \\[2mm]
&= a \cdot_\star (y \cdot_\star x) \\[2mm]
&= a \cdot_\star (x \cdot_\star y) \\[2mm]
&= (a \cdot_\star x) \cdot_\star y \\[2mm]
&= a \cdot_\star y \\[2mm]
&= b,
\end{aligned}
$$

i.e., x is a solution to the equation (1.6). This completes the proof.

Corollary 1.5. Let $a \in \mathbb{R}_\star$, $a \neq 0_\star$. Then the equation (1.5) has a unique solution.

Proof By Theorem 1.2, it follows that the equation (1.5) has at least one solution. Let x and y be two solutions to the equation (1.5). By Corollary 1.4, it follows that

$$y \cdot_\star x = y$$

and

$$x \cdot_\star y = x.$$

Hence,

$$
\begin{aligned}
y &= y \cdot_\star x \\[2mm]
&= x \cdot_\star y \\[2mm]
&= x.
\end{aligned}
$$

This completes the proof.

Remark 1.2. *By Corollary 1.5, it follows that the multiplicative unit is unique.*

Corollary 1.6. Let $a \in \mathbb{R}_\star$, $a \neq 0_\star$. Then the equation (1.4) has a unique solution.

Proof By Theorem 1.2, it follows that the equation (1.4) has at least one solution. Assume that the equation (1.4) has two solutions x and y. By Theorem 1.2, we have that the equation

$$a \cdot_\star z = 1_\star$$

has at least one solution. Then

$$
\begin{aligned}
y &= y \cdot_\star 1_\star \\[2mm]
&= y \cdot_\star (a \cdot_\star z) \\[2mm]
&= (y \cdot_\star a) \cdot_\star z \\[2mm]
&= (a \cdot_\star y) \cdot_\star z \\[2mm]
&= b \cdot_\star z \\[2mm]
&= (a \cdot_\star x) \cdot_\star z \\[2mm]
&= (x \cdot_\star a) \cdot_\star z \\[2mm]
&= x \cdot_\star (a \cdot_\star z) \\[2mm]
&= x \cdot_\star 1_\star \\[2mm]
&= x.
\end{aligned}
$$

This completes the proof.

Corollary 1.7. Let $a, b \in \mathbb{R}_\star$. Then

1. $0_\star = -_\star 0_\star$.
2. $a -_\star 0_\star = a$.
3. $a \cdot_\star 0_\star = 0_\star$.
4. $(-_\star 1_\star) \cdot_\star a = -_\star a$.
5. $(-_\star 1_\star) \cdot_\star (-_\star 1_\star) = 1_\star$.
6. $-_\star (a -_\star b) = b -_\star a$.

Proof

1. We have that $0_\star$ is a solution to the equation

$$0_\star +_\star x = 0_\star. \tag{1.7}$$

   Since

$$0_\star +_\star (-_\star 0_\star) = 0_\star,$$

   we obtain that $-_\star 0_\star$ is also a solution of the equation (1.7). By Corollary 1.2, it follows that the equation (1.7) has a unique solution. Therefore

$$0_\star = -_\star 0_\star.$$

2. We have

$$\begin{aligned}
a -_\star 0_\star &= a +_\star (-_\star 0_\star) \\
&= a +_\star 0_\star \\
&= a +_\star 1 \\
&= a \cdot 1 \\
&= a.
\end{aligned}$$

3. We have

$$\begin{aligned}
a \cdot_\star 0_\star &= a \cdot_\star 1 \\
&= e^{\log a \log 1} \\
&= 1 \\
&= 0_\star.
\end{aligned}$$

4. We have

$$\begin{aligned}
(-_\star 1_\star) \cdot_\star a &= (-_\star e) \cdot_\star a \\
&= \frac{1}{e} \cdot_\star a
\end{aligned}$$

$$= e^{\log \frac{1}{e} \log a}$$

$$= e^{-\log a}$$

$$= \frac{1}{a}$$

$$= -_\star a.$$

5. We have

$$(-_\star 1_\star) \cdot_\star (-_\star 1_\star) = (-_\star e) \cdot_\star (-_\star e)$$

$$= \frac{1}{e} \cdot_\star \frac{1}{e}$$

$$= e^{\log \frac{1}{e} \log \frac{1}{e}}$$

$$= e$$

$$= 1_\star.$$

6. We have

$$-_\star (a -_\star b) = -_\star (a +_\star (-_\star b))$$

$$= -_\star \left(a +_\star \frac{1}{b} \right)$$

$$= -_\star \frac{a}{b}$$

$$= \frac{1}{\frac{a}{b}}$$

$$= \frac{b}{a}$$

$$= b -_\star a.$$

This completes the proof.

1.2   An Order in $\mathbb{R}_\star$

Definition 1.8. We say that a number $a \in \mathbb{R}_\star$ is multiplicative positive or $\star$ positive if $a > 1$. We will write $a >_\star 0_\star$.

Example 1.5. 5 is $\star$ positive.

Definition 1.9. A number $a \in \mathbb{R}_\star$ is said to be a multiplicative negative or $\star$ negative if it is not equal to $0_\star$ and it is not $\star$ positive. We will write $a <_\star 0_\star$.

Example 1.6. $\dfrac{1}{2}$ is a $\star$ negative number.

Definition 1.10. Let $a, b \in \mathbb{R}_\star$. We say that a is multiplicative greater than b or $\star$ greater than b and we will write $a >_\star b$ if $a -_\star b >_\star 0_\star$. We will denote $a \geq_\star b$ if $a >_\star b$ or $a = b$.

Remark 1.3. *Let* $a, b \in \mathbb{R}_\star$. *Then*

$$a >_\star b \quad \Longleftrightarrow \quad a -_\star b >_\star 0_\star \quad \Longleftrightarrow \quad \frac{a}{b} > 1 \quad \Longleftrightarrow \quad a > b.$$

Definition 1.11. Let $a, b \in \mathbb{R}_\star$. We say that a is multiplicative less than b or $\star$ less than b and we will write $a <_\star b$ if $a -_\star b <_\star 0_\star$. We will denote $a \leq_\star b$ if $a <_\star b$ or $a = b$.

Remark 1.4. *Let* $a, b \in \mathbb{R}_\star$. *Then*

$$a <_\star b \quad \Longleftrightarrow \quad a -_\star b <_\star 0_\star \quad \Longleftrightarrow \quad \frac{a}{b} < 1 \quad \Longleftrightarrow \quad a < b.$$

Theorem 1.3. *Let* $a, b, c, d \in \mathbb{R}_\star$. *If*

$$a >_\star b \quad and \quad c >_\star d, \tag{1.8}$$

then

$$a +_\star c >_\star b +_\star d.$$

Proof By (1.8), it follows that

$$a > b > 0 \quad and \quad c > d > 0.$$

Then

$$
\begin{aligned}
a +_\star c &= ac \\
&> bd \\
&= b +_\star d.
\end{aligned}
$$

This completes the proof.

Theorem 1.4. *Let* $a, b, c, d \in \mathbb{R}_\star$. *If*

$$a >_\star b, \quad c >_\star d, \quad c >_\star 0_\star, \quad b >_\star 0_\star, \quad (1.9)$$

then

$$a \cdot_\star c >_\star b \cdot_\star d.$$

    Proof By (1.9), it follows that

$$a > b, \quad c > d, \quad c > 1, \quad b > 1.$$

Then

$$\log a \;>\; \log b,$$

$$\log c \;>\; \log d,$$

$$\log c \;>\; 0,$$

$$\log b \;>\; 0,$$

whereupon

$$\log a \log c \;>\; \log b \log c$$

$$>\; \log b \log d$$

and

$$e^{\log a \log c} > e^{\log b \log d},$$

i.e.,

$$a \cdot_\star c >_\star b \cdot_\star d.$$

This completes the proof.

Theorem 1.5 (Chebyschev Multiplicative Inequality). *Let* $a_1, a_2, b_1, b_2 \in \mathbb{R}_\star$. *Then the inequality*

$$a_1 \cdot_\star b_1 +_\star a_2 \cdot_\star b_2 \leq_\star e^{\frac{1}{2}} \cdot_\star (a_1 +_\star a_2) \cdot_\star (b_1 +_\star b_2)$$

if and only if

$$\log \frac{a_2}{a_1} \log \frac{b_2}{b_1} \leq 0.$$

Proof Note that

$$\log\frac{a_2}{a_1}\log\frac{b_2}{b_1}\leq 0$$

if and only if

$$\log a_1\log b_1+\log a_2\log b_2\leq\frac{1}{2}\log(a_1a_2)\log(b_1b_2)$$

if and only if

$$e^{\log a_1\log b_1+\log a_2\log b_2}\quad\leq\quad e^{\frac{1}{2}\log(a_1a_2)\log(b_1b_2)}$$

$$=\quad e^{\log e^{\frac{1}{2}}\log(a_1a_2)\log(b_1b_2)}$$

if and only if

$$a_1\cdot_\star b_1+_\star a_2\cdot_\star b_2\leq_\star e^{\frac{1}{2}}\cdot_\star(a_1+_\star a_2)\cdot_\star(b_1+_\star b_2).$$

This completes the proof.

Theorem 1.6. *Let $a,b,c,\lambda\in\mathbb{R}_\star$.*

1. *If $a>_\star b$, then $a+_\star c>_\star b+_\star c$.*
2. *If $a>_\star b$ and $b>_\star c$, then $a>_\star c$.*
3. *If $\lambda>_\star 0_\star$ and $a>_\star b$, then $\lambda\cdot_\star a>_\star\lambda\cdot_\star b$.*
4. *If $\lambda<_\star 0_\star$ and $a>_\star b$, then $\lambda\cdot_\star a<_\star\lambda\cdot_\star b$.*

Proof

1. Let $a>_\star b$. Then $a>b$ and hence,

$$ac>bc,$$

   or

$$a+_\star c>_\star b+\star c.$$

2. Let $a>_\star b$ and $b>_\star c$. Then $a>b$ and $b>c$. Hence, $a>c$ and $a>_\star c$.
3. Let $\lambda>_\star 0_\star$ and $a>_\star b$. Then $\lambda>1$ and $\log\lambda>0$, and

$$\log a>\log b.$$

   Hence,

$$\log\lambda\log a>\log\lambda\log b$$

   and

$$e^{\log\lambda\log a}>e^{\log\lambda\log b},$$

   or

$$\lambda\cdot_\star a>_\star\lambda\cdot_\star b.$$

4. Let $\lambda <_\star 0_\star$ and $a >_\star b$. Then $\lambda \in (0,1)$ and

$$\log a > \log b, \quad \log \lambda < 0.$$

Hence,
$$\log \lambda \log a < \log \lambda \log b$$

and
$$e^{\log \lambda \log a} < e^{\log \lambda \log b},$$

or
$$\lambda \cdot_\star a <_\star \lambda \cdot_\star b.$$

This completes the proof.

1.3 Multiplicative Absolute Value

In this section, we will define the multiplicative absolute value and deduct some of its properties.

Definition 1.12. Let $x \in \mathbb{R}_\star$. The multiplicative absolute value is defined as follows

$$|x|_\star = \begin{cases} x & \text{if} \quad x \geq_\star 0_\star, \\ \dfrac{1}{x} & \text{if} \quad x \leq_\star 0_\star, \end{cases}$$

or

$$|x|_\star = \begin{cases} x & \text{if} \quad x \geq 1, \\ \dfrac{1}{x} & \text{if} \quad x \leq 1. \end{cases}$$

Example 1.7. Let $x = 5$. Then $|x|_\star = 5$.

Example 1.8. Let $x = \dfrac{1}{6}$. Then $|x|_\star = 6$.

Below, we will deduct some of the properties of the multiplicative absolute value. Let $a, b \in \mathbb{R}_\star$.

1. $|a|_\star \geq_\star 0_\star$.

Proof If $a \geq_\star 0_\star$, then $a \geq 1$ and

$$|a|_\star \quad = \quad a$$

$$\geq \quad 1$$

$$\geq_\star \quad 0_\star.$$

Let $a \leq_\star 0_\star$. Then $a \leq 1$ and

$$|a|_\star \quad = \quad \frac{1}{a}$$

$$\geq \quad 1$$

$$\geq_\star \quad 0_\star.$$

This completes the proof.

2. $|a|_\star = |-_\star a|_\star$.

    Proof We have

    $$-_\star a = \frac{1}{a}.$$

    Then

    $$|a|_\star = \begin{cases} a & \text{if} \quad a \geq 1, \\ \dfrac{1}{a} & \text{if} \quad a \leq 1, \end{cases}$$

    and

    $$|-_\star a|_\star = \begin{cases} \dfrac{1}{a} & \text{if} \quad a \leq 1, \\ a & \text{if} \quad a \geq 1. \end{cases}$$

    Therefore

    $$|a|_\star = |-_\star a|_\star.$$

    This completes the proof.

3. $|a +_\star b|_\star \leq |a|_\star +_\star |b|_\star$.

    Proof We have

    $$a +_\star b = ab.$$

    We will consider the following cases:

    (a) Let $ab \geq 1$.

        i. Let $a \geq 1$, $b \leq 1$. Then

    $$|a|_\star \quad = \quad a,$$

$$|b|_\star = \frac{1}{b}$$

and

$$|a +_\star b|_\star = |ab|_\star$$

$$= ab$$

$$\leq a\frac{1}{b}$$

$$= a +_\star \frac{1}{b}$$

$$= |a|_\star +_\star |b|_\star.$$

    ii. The case $a \leq 1$ and $b \geq 1$ we leave to the reader as an exercise.

  iii. Let $a \geq 1$ and $b \geq 1$. Then

$$|a|_\star = a,$$

$$|b|_\star = b.$$

Hence,

$$|a +_\star b|_\star = |ab|_\star$$

$$= ab$$

$$= a +_\star b$$

$$= |a|_\star +_\star |b|_\star.$$

(b) Let $ab \leq 1$.

    i. Let $a \leq 1$ and $b \leq 1$. Then

$$|a|_\star = \frac{1}{a},$$

$$|b|_\star = \frac{1}{b},$$

$$|ab|_\star = \frac{1}{ab}.$$

Hence,

$$
\begin{aligned}
|a +_\star b|_\star &= |ab|_\star \\
&= \frac{1}{ab} \\
&= |a|_\star |b|_\star \\
&= |a|_\star +_\star |b|_\star.
\end{aligned}
$$

ii. Let $a \geq 1, b \leq 1$. Then

$$
\begin{aligned}
|a|_\star &= a, \\
|b|_\star &= \frac{1}{b}, \\
|ab|_\star &= \frac{1}{ab}, \\
\frac{1}{a} &\leq 1.
\end{aligned}
$$

Hence,

$$
\begin{aligned}
|a +_\star b|_\star &= |ab|_\star \\
&= \frac{1}{ab} \\
&\leq \frac{a}{b} \\
&= |a|_\star |b|_\star \\
&= |a|_\star +_\star |b|_\star.
\end{aligned}
$$

iii. The case $a \leq 1$ and $b \geq 1$ we leave to the reader as an exercise. This completes the proof.

4. $|a -_\star b|_\star \geq_\star |a|_\star -_\star |b|_\star.$

Proof We have

$$
|a|_\star -_\star |b|_\star = |a -_\star b +_\star b|_\star -_\star |b|_\star
$$

$$\leq \quad |a-_\star b|_\star +_\star |b|_\star -_\star |b|_\star$$

$$= \quad |a-_\star b|_\star.$$

This completes the proof.

Exercise 1.3. Let $a, b \in \mathbb{R}_\star$. Prove that

1. $||a|_\star -_\star |b|_\star|_\star \leq_\star |a -_\star b|_\star$.
2. $|a \cdot_\star b|_\star = |a|_\star \cdot_\star |b|_\star$.
3. $|a/_\star b|_\star = |a|_\star /_\star |b|_\star$.

1.4 The Multiplicative Factorial. Multiplicative Binomial Coefficients

Definition 1.13. For $n \in \mathbb{N}$, define the multiplicative factorial $n!_\star$ as follows

$$n!_\star = e^{n!}.$$

Example 1.9. We have

$$3!_\star = e^{3!}$$

$$= e^{3 \cdot 2}$$

$$= e^6.$$

Example 1.10. We have

$$4!_\star = e^{4!}$$

$$= e^{4 \cdot 3 \cdot 2}$$

$$= e^{24}.$$

Example 1.11. We have

$$5!_\star = e^{5!}$$

$$= e^{5 \cdot 4 \cdot 3 \cdot 2}$$

$$= e^{120}.$$

Exercise 1.4. Compute $6!_\star$.

Answer 1.3. e^{720}.

Definition 1.14. For $n, k \in \mathbb{N}$, $n > k$, define

$$\left(\begin{array}{c} n \\ k \end{array} \right)_\star = e^{\left(\begin{array}{c} n \\ k \end{array} \right)}.$$

Example 1.12. We have

$$\left(\begin{array}{c} 4 \\ 2 \end{array} \right)_\star = e^{\left(\begin{array}{c} 4 \\ 2 \end{array} \right)}$$

$$= e^{\frac{4!}{2!2!}}$$

$$= e^{\frac{24}{2 \cdot 2}}$$

$$= e^6.$$

Exercise 1.5. Find

$$\left(\begin{array}{c} 5 \\ 2 \end{array} \right)_\star.$$

Answer 1.4. e^{10}.

Theorem 1.7. *For $n, k \in \mathbb{N}$, $n > k$, we have*

$$\left(\begin{array}{c} n \\ k \end{array} \right)_\star = \left(\begin{array}{c} n-1 \\ k \end{array} \right)_\star +_\star \left(\begin{array}{c} n-1 \\ k-1 \end{array} \right)_\star.$$

Proof We have

$$\left(\begin{array}{c} n \\ k \end{array} \right)_\star = e^{\left(\begin{array}{c} n \\ k \end{array} \right)}$$

$$= e^{\left(\begin{array}{c} n-1 \\ k \end{array} \right) + \left(\begin{array}{c} n-1 \\ k-1 \end{array} \right)}$$

$$= e^{\left(\begin{array}{c} n \\ k \end{array}\right)}$$

$$= e^{\left(\begin{array}{c} n-1 \\ k \end{array}\right)} \cdot e^{\left(\begin{array}{c} n-1 \\ k-1 \end{array}\right)}$$

$$= e^{\left(\begin{array}{c} n \\ k \end{array}\right)}$$

$$= e^{\left(\begin{array}{c} n-1 \\ k \end{array}\right)} +_\star e^{\left(\begin{array}{c} n-1 \\ k-1 \end{array}\right)}$$

$$= \left(\begin{array}{c} n-1 \\ k \end{array}\right)_\star +_\star \left(\begin{array}{c} n-1 \\ k-1 \end{array}\right)_\star.$$

This completes the proof.

1.5   Multiplicative Functions of One Variable

Definition 1.15. For a function $f : \mathbb{R}_\star \to \mathbb{R}$, we define the function $f_\star : \mathbb{R}_\star \to [0, \infty)$ as follows

$$f_\star(x) = e^{f(\log x)}, \quad x \in \mathbb{R}_\star.$$

Example 1.13. Let $f(x) = 1 - x$, $x \in \mathbb{R}_\star$. Then $f_\star(x) = e^{1 - \log x}$, $x \in \mathbb{R}_\star$. In Fig. 1.1, it is shown the graphics of f and $f_\star$ on $[5, 10]$.

Example 1.14. Let

$$f(x) = x^2, \quad g(x) = x^3, \quad x \in \mathbb{R}_\star.$$

We will find

$$f_\star(x) +_\star g_\star(x), \quad x \in \mathbb{R}_\star.$$

We have

$$f_\star(x) = e^{(\log x)^2},$$

$$g_\star(x) = e^{(\log x)^3}, \quad x \in \mathbb{R}_\star.$$

Hence,

$$f_\star(x) +_\star g_\star(x) = e^{(\log x)^2} +_\star e^{(\log x)^3}$$

$$= e^{(\log x)^2 + (\log x)^3}, \quad x \in \mathbb{R}_\star.$$

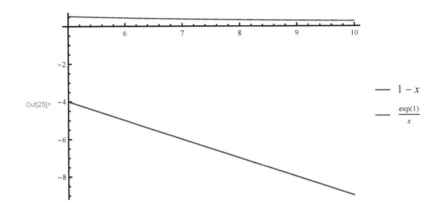

FIGURE 1.1
The graphics of the functions $f(x) = 1 - x$, $x \in \mathbb{R}_\star$, and $f_\star(x) = e^{1-\log x}$, $x \in \mathbb{R}_\star$.

Exercise 1.6. Let
$$f(x) = x, \quad g(x) = x^2, \quad x \in \mathbb{R}_\star.$$
Find

      1. $f_\star(x)$, $x \in \mathbb{R}_\star$.

      2. $g_\star(x)$, $x \in \mathbb{R}_\star$.

      3. $f_\star(x) +_\star g_\star(x)$, $x \in \mathbb{R}_\star$.

      4. $f_\star(x) -_\star g_\star(x)$, $x \in \mathbb{R}_\star$.

      5. $f_\star(x) \cdot_\star g_\star(x)$, $x \in \mathbb{R}_\star$.

      6. $f_\star(x)/_\star g_\star(x)$, $x \in \mathbb{R}_\star$.

Answer 1.5.   *1. x, $x \in \mathbb{R}_\star$.*

      2. $e^{(\log x)^2}$, $x \in \mathbb{R}_\star$.

      3. $xe^{(\log x)^2}$, $x \in \mathbb{R}_\star$.

      4. $xe^{-(\log x)^2}$, $x \in \mathbb{R}_\star$.

      5. $e^{(\log x)^3}$, $x \in \mathbb{R}_\star$.

      6. $e^{\frac{1}{\log x}}$, $x \in \mathbb{R}_\star$.

Definition 1.16. Let $A \subseteq \mathbb{R}_\star$. The function $f_\star : A \to [0, \infty)$ is said to be a continuous multiplicative function at the point $x_0 \in A$ if $e^{f(\log x)}$ is continuous at x_0. The function

$f_\star$ is said to be a continuous multiplicative function on A if it is continuous at any point of A.

Since the logarithm function and the exponential function are continuous functions on $\mathbb{R}_\star$, by the above definition, it follows that if f is a continuous function on A, then $f_\star$ is a continuous function on A and the conversely.

1.6   The Multiplicative Power Function

In this section, we will define the power function and we will deduct some of its properties.

Definition 1.17. Let $x \in \mathbb{R}_\star$ and $k \in \mathbb{N}$. Define

$$x^{k_\star} = \underbrace{x \cdot_\star \cdot_\star \cdots \cdot_\star x}_{k}.$$

By Definition 1.17, it follows

$$
\begin{aligned}
x^{2_\star} &= x \cdot_\star x \\
&= e^{\log x \log x} \\
&= e^{(\log x)^2}.
\end{aligned}
$$

Assume that

$$x^{k_\star} = e^{(\log x)^k}$$

for some $k \in \mathbb{N}$. We will prove that

$$x^{k+1_\star} = e^{(\log x)^{k+1}}.$$

Really,

$$
\begin{aligned}
x^{k+1_\star} &= \underbrace{x \cdot_\star \cdots \cdot_\star x}_{k} \cdot_\star x \\
&= e^{(\log x)^k} \cdot_\star x \\
&= e^{\log e^{(\log x)^k} \log x} \\
&= e^{(\log x)^k \log x} \\
&= e^{(\log x)^{k+1}}.
\end{aligned}
$$

Example 1.15. We have
$$2^{3_\star} = e^{(\log 2)^3}.$$

Example 1.16. We will compute
$$A = (3 +_\star 4) \cdot_\star 3^{2_\star}.$$

We have
$$3 +_\star 4 \;=\; 3 \cdot 4$$
$$=\; 12,$$
$$3^{2_\star} \;=\; e^{(\log 3)^2}.$$

Then
$$A \;=\; 12 \cdot_\star e^{(\log 3)^2}$$
$$=\; e^{(\log(12))(\log 3)^2}$$
$$=\; e^{(2\log 2 + \log 3)(\log 3)^2}.$$

Exercise 1.7. Compute

1. $A = ((2 -_\star 1) /_\star 3)^{2_\star}$.
2. $A = ((4 +_\star 2) \cdot_\star 2)^{2_\star} -_\star 2$.
3. $A = ((5 -_\star 3) /_\star 4)^{2_\star} +_\star (3 \cdot_\star 2)$.

Answer 1.6.   *1.* $e^{\left(\frac{\log 2}{\log 3}\right)^2}$.

2. $\dfrac{1}{2} e^{9(\log 2)^4}$.

3. $e^{\log 3 \log 2 + \frac{\left(\log \frac{5}{3}\right)^2}{4(\log 2)^2}}$.

Definition 1.18. Let $x \in \mathbb{R}_\star$ and $k \in \mathbb{N}$. Define
$$x^{-k_\star} = e^{\left(\frac{1}{\log x}\right)^k}.$$

Note that
$$x^{-k_\star} \cdot_\star x^{k_\star} \;=\; e^{\left(\frac{1}{\log x}\right)^k} \cdot_\star e^{(\log x)^k}$$

$$= \quad e^{\log e \left(\frac{1}{\log x} \right)^k \log e^{(\log x)^k}}$$

$$= \quad e^{\left(\frac{1}{\log x} \right)^k (\log x)^k}$$

$$= \quad e$$

$$= \quad 1_\star.$$

Example 1.17. We have

$$5^{-4_\star} = e^{\left(\frac{1}{\log 5} \right)^4}.$$

Definition 1.19. Let $x \in \mathbb{R}_\star$, $p, q \in \mathbb{N}$. Define

$$x^{\frac{p}{q}_\star} = e^{(\log x)^{\frac{p}{q}}}.$$

Note that

$$\left(x^{\frac{p}{q}_\star} \right)^{q_\star} = \left(e^{(\log x)^{\frac{p}{q}}} \right)^{q_\star}$$

$$= e^{\left(\log e^{(\log x)^{\frac{p}{q}}} \right)^q}$$

$$= e^{\left((\log x)^{\frac{p}{q}} \right)^q}$$

$$= e^{(\log x)^p}$$

$$= x^{p_\star}.$$

Example 1.18. We have

$$2^{\frac{3}{4}_\star} = e^{(\log 2)^{\frac{3}{4}}}.$$

Definition 1.20. Let $x \in \mathbb{R}_\star$ and $k \in \mathbb{R}$. Define

$$x^{k_\star} = e^{(\log x)^k}.$$

Note that for any $k \in \mathbb{R}$, we have

$$1_\star^{k_\star} = e^{(\log e)^k}$$

$$= e$$

$$= 1_\star.$$

Now, we will deduct some useful formulae.

1. $(x+_\star y)^{2_\star} = x^{2_\star} +_\star e^2 \cdot_\star x \cdot_\star y +_\star y^{2_\star}, x, y \in \mathbb{R}_\star.$

   Proof We have

   $$
   \begin{aligned}
   (x+_\star y)^{2_\star} &= (xy)^{2_\star} \\[2mm]
   &= e^{(\log(xy))^2} \\[2mm]
   &= e^{(\log x + \log y)^2} \\[2mm]
   &= e^{(\log x)^2 + 2\log x \log y + (\log y)^2} \\[2mm]
   &= e^{(\log x)^2} e^{2\log x \log y} e^{(\log y)^2} \\[2mm]
   &= e^{(\log x)^2} +_\star e^{2\log x \log y} +_\star e^{(\log y)^2} \\[2mm]
   &= x^{2_\star} +_\star e^2 \cdot_\star e^{\log x \log y} +_\star e^{(\log y)^2} \\[2mm]
   &= x^{2_\star} +_\star e^2 \cdot_\star x \cdot_\star y +_\star y^{2_\star}.
   \end{aligned}
   $$

   This completes the proof.

2. $(x-_\star y)^{2_\star} = x^{2_\star} -_\star e^2 \cdot_\star x \cdot_\star y +_\star y^{2_\star}, x, y \in \mathbb{R}_\star.$

   Proof We have

   $$
   \begin{aligned}
   (x-_\star y)^{2_\star} &= \left(\frac{x}{y}\right)^{2_\star} \\[2mm]
   &= e^{(\log \frac{x}{y})^2} \\[2mm]
   &= e^{(\log x - \log y)^2} \\[2mm]
   &= e^{(\log x)^2 - 2\log x \log y + (\log y)^2} \\[2mm]
   &= e^{(\log x)^2} e^{-2\log x \log y} e^{(\log y)^2} \\[2mm]
   &= e^{(\log x)^2} -_\star e^{2\log x \log y} +_\star e^{(\log y)^2}
   \end{aligned}
   $$

$$= \quad x^{2\star} -_\star e^2 \cdot_\star e^{\log x \log y} +_\star e^{(\log y)^2}$$

$$= \quad x^{2\star} -_\star e^2 \cdot_\star x \cdot_\star y +_\star y^{2\star}.$$

This completes the proof.

3. $(x +_\star y)^{3\star} = x^{3\star} +_\star e^3 \cdot_\star x^{2\star} \cdot_\star y +_\star e^3 \cdot_\star x \cdot_\star y^{2\star} +_\star y^{3\star}$, $x, y \in \mathbb{R}_\star$.

Proof We have

$$(x +_\star y)^{3\star} \quad = \quad (xy)^{3\star}$$

$$= \quad e^{(\log(xy))^3}$$

$$= \quad e^{(\log x + \log y)^3}$$

$$= \quad e^{(\log x)^3 + 3(\log x)^2 \log y + 3 \log x (\log y)^2 + (\log y)^3}$$

$$= \quad e^{(\log x)^3} +_\star e^{3(\log x)^2 \log y} +_\star e^{3 \log x (\log y)^2} +_\star e^{(\log y)^3}$$

$$= \quad x^{3\star} +_\star e^3 \cdot_\star e^{(\log x)^2} \cdot_\star e^{\log y} +_\star e^3 \cdot_\star e^{\log x} \cdot_\star e^{(\log y)^2}$$

$$+_\star e^{(\log y)^3}$$

$$= \quad x^{3\star} +_\star e^3 \cdot_\star x^{2\star} \cdot_\star y +_\star e^3 \cdot_\star x \cdot_\star y^{2\star} +_\star y^{3\star}.$$

This completes the proof.

4. $(x -_\star y)^{3\star} = x^{3\star} -_\star e^3 \cdot_\star x^{2\star} \cdot_\star y +_\star e^3 \cdot_\star x \cdot_\star y^{2\star} -_\star y^{3\star}$, $x, y \in \mathbb{R}_\star$.

Proof We have

$$(x -_\star y)^{3\star} \quad = \quad \left(\frac{x}{y}\right)^{3\star}$$

$$= \quad e^{(\log \frac{x}{y})^3}$$

$$= \quad e^{(\log x - \log y)^3}$$

$$= \quad e^{(\log x)^3 - 3(\log x)^2 \log y + 3 \log x (\log y)^2 - (\log y)^3}$$

$$= \quad e^{(\log x)^3} -_\star e^{3(\log x)^2 \log y} +_\star e^{3\log x (\log y)^2} -_\star e^{(\log y)^3}$$

$$= \quad x^{3\star} -_\star e^3 \cdot_\star e^{(\log x)^2} \cdot_\star e^{\log y} +_\star e^3 \cdot_\star e^{\log x} \cdot_\star e^{(\log y)^2}$$

$$-_\star e^{(\log y)^3}$$

$$= \quad x^{3\star} -_\star e^3 \cdot_\star x^{2\star} \cdot_\star y +_\star e^3 \cdot_\star x \cdot_\star y^{2\star} -_\star y^{3\star}.$$

This completes the proof.

Exercise 1.8. Prove

   1. $(x +_\star y) \cdot_\star (x -_\star y) = x^{2\star} -_\star y^{2\star}$, $x, y \in \mathbb{R}_\star$.

   2. $(x +_\star y) \cdot_\star (x^{2\star} -_\star x \cdot_\star y +_\star y^{2\star}) = x^{3\star} +_\star y^{3\star}$, $x, y \in \mathbb{R}_\star$.

   3. $(x -_\star y) \cdot_\star (x^{2\star} -_\star x \cdot_\star y +_\star y^{2\star}) = x^{3\star} -_\star y^{3\star}$, $x, y \in \mathbb{R}_\star$.

Definition 1.21. For $k \in \mathbb{R}$ and $f : \mathbb{R}_\star \to [0, \infty)$. define

$$(f_\star(x))^{k_\star} = e^{(f(x))^k}, \quad x \in \mathbb{R}_\star.$$

Example 1.19. Let $f(x) = \sin x$, $x \in \mathbb{R}$. Then

$$(f_\star(x))^{\frac{1}{3}\star} \quad = \quad e^{(f(\log x))^{\frac{1}{3}}}$$

$$= \quad e^{(\sin(\log x))^{\frac{1}{3}}}, \quad x \in \mathbb{R}_\star.$$

Definition 1.22. For $a \in \mathbb{R}_\star$, define

$$a_\star^x = e^{a^{\log x}}, \quad x \in \mathbb{R}_\star.$$

In particular, we have

$$e_\star^x \quad = \quad e^{e^{\log x}}$$

$$= \quad e^x, \quad x \in \mathbb{R}_\star.$$

Below, we will deduct some of the properties of the function $a_\star^x$, $x \in \mathbb{R}_\star$.

   1. $a_\star^x \cdot_\star a_\star^y = a_\star^{x +_\star y}$, $x, y \in \mathbb{R}_\star$.

     Proof We have

$$a_\star^x \cdot_\star a_\star^y \quad = \quad e^{a^{\log x}} \cdot_\star e^{a^{\log y}}$$

$$= e^{a^{\log x} a^{\log y}}$$

$$= e^{a^{\log x + \log y}}$$

$$= e^{a^{\log(xy)}}$$

$$= a_\star^{xy}$$

$$= a_\star^{x +_\star y}, \quad x, y \in \mathbb{R}_\star.$$

This completes the proof.

2. $a_\star^x /_\star a_\star^y = a_\star^{x -_\star y}$, $x, y \in \mathbb{R}_\star$.

   Proof We have

$$a_\star^x /_\star a_\star^y = e^{a^{\log x}} /_\star e^{a^{\log y}}$$

$$= e^{a^{\frac{\log x}{\log y}}}$$

$$= e^{a^{\log x - \log y}}$$

$$= e^{a^{\log \frac{x}{y}}}$$

$$= a_\star^{\frac{x}{y}}$$

$$= a_\star^{x -_\star y}, \quad x, y \in \mathbb{R}_\star.$$

This completes the proof.

1.7 Multiplicative Trigonometric Functions

Definition 1.23. For $x \in \mathbb{R}_\star$, define multiplicative sine, multiplicative cosine, multiplicative tangent and multiplicative cotangent as follows

1. $\sin_\star x = e^{\sin \log x}$.
2. $\cos_\star x = e^{\cos \log x}$.
3. $\tan_\star x = e^{\tan \log x}$.
4. $\cot_\star x = e^{\cot \log x}$.

Example 1.20. We have

$$\sin_\star e^{\frac{\pi}{2}} \quad = \quad e^{\sin \log e^{\frac{\pi}{2}}}$$

$$= \quad e^{\sin \frac{\pi}{2}}$$

$$= \quad e^1$$

$$= \quad 1_\star$$

and

$$\cos_\star e^{\frac{\pi}{2}} \quad = \quad e^{\cos \log e^{\frac{\pi}{2}}}$$

$$= \quad e^{\cos \frac{\pi}{2}}$$

$$= \quad 1$$

$$= \quad 0_\star.$$

Example 1.21. We will compute

$$A = 2 \cdot_\star \sin_\star e^{\frac{\pi}{4}} +_\star \cos_\star e^{\frac{\pi}{4}}.$$

We have

$$\sin_\star e^{\frac{\pi}{4}} \quad = \quad e^{\sin \log e^{\frac{\pi}{4}}}$$

$$= \quad e^{\sin \frac{\pi}{4}}$$

$$= \quad e^{\frac{\sqrt{2}}{2}},$$

$$2 \cdot_\star \sin_\star e^{\frac{\pi}{4}} \quad = \quad e^{\log 2 \log e^{\frac{\sqrt{2}}{2}}}$$

$$= \quad e^{\frac{\sqrt{2} \log 2}{2}},$$

$$\cos_\star e^{\frac{\pi}{4}} \quad = \quad e^{\cos \left(\log e^{\frac{\pi}{4}} \right)}$$

$$= \quad e^{\cos \frac{\pi}{4}}$$

$$= e^{\frac{\sqrt{2}}{2}}.$$

Therefore

$$A = e^{\frac{\sqrt{2}\log 2}{2}} +_\star e^{\frac{\sqrt{2}}{2}}$$

$$= e^{\frac{\sqrt{2}}{2}\log 2} e^{\frac{\sqrt{2}}{2}}$$

$$= e^{\frac{\sqrt{2}}{2}(1+\log 2)}.$$

Exercise 1.9. Find

$$A = 3 \cdot_\star \cos_\star e^{\frac{\pi}{4}} -_\star \sin_\star e^{\frac{\pi}{2}} +_\star 2 \cdot_\star \cos_\star e^{\frac{\pi}{2}} +_\star \cot_\star e^{\frac{\pi}{4}}.$$

Answer 1.7. $e^{\frac{\sqrt{2}}{2}\log 3}$.

We will deduct some of the basic properties of the multiplicative trigonometric functions.

1. $\sin_\star x +_\star \sin_\star y = e^2 \cdot_\star \sin_\star \left((x +_\star y)/_\star e^2\right) \cdot_\star \cos_\star \left((x -_\star y)/_\star e^2\right).$

    Proof We have

    $$\sin_\star x +_\star \sin_\star y = e^{\sin\log x} +_\star e^{\sin\log y}$$

    $$= e^{\sin\log x + \sin\log y}$$

    $$= e^{2\sin\frac{\log x + \log y}{2}\cos\frac{\log x - \log y}{2}}$$

    $$= e^{2\sin\left(\log\sqrt{xy}\right)\cos\left(\log\sqrt{\frac{x}{y}}\right)}$$

    $$= e^2 \cdot_\star e^{\sin\left(\log\sqrt{xy}\right)\cos\left(\log\sqrt{\frac{x}{y}}\right)}$$

    $$= e^2 \cdot_\star e^{\sin\left(\log\sqrt{xy}\right)} \cdot_\star e^{\cos\left(\log\sqrt{\frac{x}{y}}\right)}$$

    $$= e^2 \cdot_\star e^{\sin\left(\log((x+_\star y)/_\star e^2)\right)} \cdot_\star e^{\cos\left(\log((x-_\star y)/_\star e^2)\right)}$$

    $$= e^2 \cdot_\star \sin_\star \left((x +_\star y)/_\star e^2\right) \cdot_\star \cos_\star \left((x -_\star y)/_\star e^2\right).$$

    This completes the proof.

2. $\sin_\star x -_\star \sin_\star y = e^2 \cdot_\star \sin_\star \left((x -_\star y) /_\star e^2 \right) \cdot_\star \cos_\star \left((x +_\star y) /_\star e^2 \right).$

Proof We have

$$\sin_\star x -_\star \sin_\star y = e^{\sin\log x} -_\star e^{\sin\log y}$$

$$= e^{\sin\log x - \sin\log y}$$

$$= e^{2\sin\frac{\log x - \log y}{2}\cos\frac{\log x + \log y}{2}}$$

$$= e^{2\cos\left(\log\sqrt{xy}\right)\sin\left(\log\sqrt{\frac{x}{y}}\right)}$$

$$= e^2 \cdot_\star e^{\cos\left(\log\sqrt{xy}\right)\sin\left(\log\sqrt{\frac{x}{y}}\right)}$$

$$= e^2 \cdot_\star e^{\cos\left(\log\sqrt{xy}\right)} \cdot_\star e^{\sin\left(\log\sqrt{\frac{x}{y}}\right)}$$

$$= e^2 \cdot_\star e^{\cos\left(\log\left((x+_\star y)/_\star e^2\right)\right)} \cdot_\star e^{\sin\left(\log\left((x-_\star y)/_\star e^2\right)\right)}$$

$$= e^2 \cdot_\star \sin_\star \left((x -_\star y) /_\star e^2 \right) \cdot_\star \cos_\star \left((x +_\star y) /_\star e^2 \right).$$

This completes the proof.

3. $(\sin_\star x)^{2_\star} +_\star (\cos_\star x)^{2_\star} = 1_\star.$

Proof We have

$$(\sin_\star x)^{2_\star} +_\star (\cos_\star x)^{2_\star} = \left(e^{\sin\log x}\right)^{2_\star} +_\star \left(e^{\cos\log x}\right)^{2_\star}$$

$$= e^{\left(\log e^{\sin\log x}\right)^2} +_\star e^{\left(\log e^{\cos\log x}\right)^2}$$

$$= e^{(\sin\log x)^2} +_\star e^{(\cos\log x)^2}$$

$$= e^{(\sin\log x)^2 + (\cos\log x)^2}$$

$$= e$$

$$= 1_\star.$$

This completes the proof.

4. $(\sin_\star x)^{2\star} = \left(1_\star -_\star \cos_\star \left(e^2 \cdot_\star x\right)\right) /_\star e^2.$

Proof We have

$$
\left(1_\star -_\star \cos_\star \left(e^2 \cdot_\star x\right)\right) /_\star e^2 = \left(e -_\star e^{\cos\left(\log\left(e^2 \cdot_\star x\right)\right)}\right) /_\star e^2
$$

$$
= \left(e -_\star e^{\cos\left(\log e^{2\log x}\right)}\right) /_\star e^2
$$

$$
= \left(e -_\star e^{\cos(2\log x)}\right) /_\star e^2
$$

$$
= \left(\frac{e}{e^{\cos(2\log x)}}\right) /_\star e^2
$$

$$
= \left(e^{1-\cos(2\log x)}\right) /_\star e^2
$$

$$
= \left(e^{2(\sin\log x)^2}\right) /_\star e^2
$$

$$
= e^{\frac{\log e^{2(\sin\log x)^2}}{\log e^2}}
$$

$$
= e^{(\sin\log x)^2}
$$

$$
= e^{\left(\log e^{\sin\log x}\right)^2}
$$

$$
= \left(e^{\sin\log x}\right)^{2\star}
$$

$$
= (\sin_\star x)^{2\star}.
$$

This completes the proof.

5. $(\cos_\star x)^{2\star} = \left(1_\star +_\star \cos_\star \left(e^2 \cdot_\star x\right)\right) /_\star e^2.$

Proof We have

$$
\left(1_\star +_\star \cos_\star \left(e^2 \cdot_\star x\right)\right) /_\star e^2 = \left(e +_\star e^{\cos\left(\log\left(e^2 \cdot_\star x\right)\right)}\right) /_\star e^2
$$

$$
= \left(e +_\star e^{\cos\left(\log e^{2\log x}\right)}\right) /_\star e^2
$$

$$
= \left(e +_\star e^{\cos(2\log x)}\right) /_\star e^2
$$

$$
= \left(e^{1+\cos(2\log x)}\right) /_\star e^2
$$

$$= \left(e^{2(\cos\log x)^2}\right)/_\star e^2$$

$$= e^{\frac{\log e^{2(\cos\log x)^2}}{\log e^2}}$$

$$= e^{(\cos\log x)^2}$$

$$= e^{\left(\log e^{\cos\log x}\right)^2}$$

$$= \left(e^{\cos\log x}\right)^{2_\star}$$

$$= (\cos_\star x)^{2_\star}.$$

This completes the proof.

6. $\cos_\star x +_\star \cos_\star y = e^2 \cdot_\star \cos_\star \left((x +_\star y)/_\star e^2\right) \cdot_\star \cos_\star \left((x -_\star y)/_\star e^2\right).$

Proof We have

$$\cos_\star x +_\star \cos_\star y = e^{\cos\log x} +_\star e^{\cos\log y}$$

$$= e^{\cos\log x + \cos\log y}$$

$$= e^{2\cos\frac{\log x + \log y}{2} \cos\frac{\log x - \log y}{2}}$$

$$= e^{2\cos\left(\log\sqrt{xy}\right)\cos\left(\log\sqrt{\frac{x}{y}}\right)}$$

$$= e^2 \cdot_\star e^{\cos\left(\log\sqrt{xy}\right)\cos\left(\log\sqrt{\frac{x}{y}}\right)}$$

$$= e^2 \cdot_\star e^{\cos\left(\log\sqrt{xy}\right)} \cdot_\star e^{\cos\left(\log\sqrt{\frac{x}{y}}\right)}$$

$$= e^2 \cdot_\star e^{\cos\left(\log\left((x+_\star y)/_\star e^2\right)\right)} \cdot_\star e^{\cos\left(\log\left((x-_\star y)/_\star e^2\right)\right)}$$

$$= e^2 \cdot_\star \cos_\star \left((x +_\star y)/_\star e^2\right) \cdot_\star \cos_\star \left((x -_\star y)/_\star e^2\right).$$

This completes the proof.

7. $\cos_\star x -_\star \cos_\star y = -_\star e^2 \cdot_\star \sin_\star \left((x -_\star y)/_\star e^2\right) \cdot_\star \sin_\star \left((x +_\star y)/_\star e^2\right).$

Proof We have

$$\cos_\star x -_\star \cos_\star y = e^{\cos\log x} -_\star e^{\cos\log y}$$

$$= \quad e^{\cos\log x - \cos\log y}$$

$$= \quad e^{-2\sin\frac{\log x - \log y}{2}\,\sin\frac{\log x + \log y}{2}}$$

$$= \quad e^{-2\sin\left(\log\sqrt{xy}\right)\sin\left(\log\sqrt{\frac{x}{y}}\right)}$$

$$= \quad -_\star e^2 \cdot_\star e^{\sin\left(\log\sqrt{xy}\right)\sin\left(\log\sqrt{\frac{x}{y}}\right)}$$

$$= \quad -_\star e^2 \cdot_\star e^{\sin\left(\log\sqrt{xy}\right)} \cdot_\star e^{\sin\left(\log\sqrt{\frac{x}{y}}\right)}$$

$$= \quad -_\star e^2 \cdot_\star e^{\sin\left(\log\left((x+_\star y)/_\star e^2\right)\right)} \cdot_\star e^{\sin\left(\log\left((x-_\star y)/_\star e^2\right)\right)}$$

$$= \quad -_\star e^2 \cdot_\star \sin_\star\left((x-_\star y)/_\star e^2\right) \cdot_\star \sin_\star\left((x+_\star y)/_\star e^2\right).$$

This completes the proof.

8. $\tan_\star x = \sin_\star x/_\star \cos_\star x.$

Proof We have

$$\tan_\star x \quad = \quad e^{\tan\log x}$$

$$= \quad e^{\frac{\sin\log x}{\cos\log x}}$$

$$= \quad e^{\frac{\log e^{\sin\log x}}{\log e^{\cos\log x}}}$$

$$= \quad \left(e^{\sin\log x}\right)/_\star \left(e^{\cos\log x}\right)$$

$$= \quad \sin_\star x/_\star \cos_\star x.$$

This completes the proof.

9. $\cot_\star x = \cos_\star x/_\star \sin_\star x.$

Proof We have

$$\cot_\star x \quad = \quad e^{\cot\log x}$$

$$= e^{\frac{\cos\log x}{\sin\log x}}$$

$$= e^{\frac{\log e^{\cos\log x}}{\log e^{\sin\log x}}}$$

$$= \left(e^{\cos\log x}\right)/_\star\left(e^{\sin\log x}\right)$$

$$= \cos_\star x/_\star \sin_\star x.$$

This completes the proof.

1.8 Multiplicative Inverse Trigonometric Functions

In this section, we will define the multiplicative inverse trigonometric functions.

Definition 1.24. For $x \in \mathbb{R}_\star$, we define the multiplicative inverse trigonometric functions as follows

1. $\arcsin_\star x = e^{\arcsin(\log x)}$.
2. $\arccos_\star x = e^{\arccos(\log x)}$.
3. $\arctan_\star x = e^{\arctan(\log x)}$.
4. $\mathrm{arccot}_\star x = e^{\mathrm{arccot}(\log x)}$.

Example 1.22. We have

$$\arcsin_\star e^{\frac{\sqrt{2}}{2}} = e^{\arcsin\left(\log e^{\frac{\sqrt{2}}{2}}\right)}$$

$$= e^{\arcsin\left(\frac{\sqrt{2}}{2}\right)}$$

$$= e^{\frac{\pi}{4}}$$

and

$$\arctan_\star e = e^{\arctan\log e}$$

$$= e^{\arctan 1}$$

$$= e^{\frac{\pi}{4}}.$$

Now, we have listed some of the properties of the multiplicative inverse trigonometric functions.

1. For $x \in \mathbb{R}_\star$, we have

$$\sin_\star (\arcsin_\star x) = e^{\sin(\log(\arcsin_\star x))}$$

$$= e^{\sin\left(\log\left(e^{\arcsin(\log x)}\right)\right)}$$

$$= e^{\sin(\arcsin(\log x))}$$

$$= e^{\log x}$$

$$= x.$$

2. For $x \in \mathbb{R}_\star$, we have

$$\arcsin_\star (\sin_\star x) = e^{\arcsin(\log(\sin_\star x))}$$

$$= e^{\arcsin\left(\log\left(e^{\sin(\log x)}\right)\right)}$$

$$= e^{\arcsin(\sin(\log x))}$$

$$= e^{\log x}$$

$$= x.$$

3. For $x \in \mathbb{R}_\star$, we have

$$\cos_\star (\arccos_\star x) = e^{\cos(\log(\arccos_\star x))}$$

$$= e^{\cos\left(\log e^{\arccos(\log x)}\right)}$$

$$= e^{\cos(\arccos(\log x))}$$

$$= e^{\log x}$$

$$= x.$$

Exercise 1.10. Prove that

$$\arccos_\star (\cos_\star x) = x, \quad x \in \mathbb{R}_\star.$$

Example 1.23. We will find

$$\arcsin_\star (\cos_\star x), \quad x \in \mathbb{R}_\star.$$

We have

$$
\begin{aligned}
\arcsin_\star (\cos_\star x) &= e^{\arcsin(\log(\cos_\star x))} \\[2mm]
&= e^{\arcsin\left(\log\left(e^{\cos(\log x)}\right)\right)} \\[2mm]
&= e^{\arcsin(\cos(\log x))} \\[2mm]
&= e^{\arcsin\left(\sin\left(\frac{\pi}{2}-\log x\right)\right)} \\[2mm]
&= e^{\frac{\pi}{2}-\log x} \\[2mm]
&= \frac{e^{\frac{\pi}{2}}}{x} \\[2mm]
&= e^{\frac{\pi}{2}} -_\star x,, \quad x \in \mathbb{R}_\star.
\end{aligned}
$$

4. For $x \in \mathbb{R}_\star$, we have

$$
\begin{aligned}
\tan_\star (\arctan_\star x) &= e^{\tan(\log(\arctan_\star x))} \\[2mm]
&= e^{\tan\left(\log e^{\arctan(\log x)}\right)} \\[2mm]
&= e^{\tan(\arctan(\log x))} \\[2mm]
&= e^{\log x} \\[2mm]
&= x.
\end{aligned}
$$

Exercise 1.11. Prove that

$$\arctan_\star (\tan_\star x) = x, \quad x \in \mathbb{R}_\star.$$

5. For $x \in \mathbb{R}_\star$, we have

$$\cos_\star (\arccos_\star x) = e^{\cos(\log(\arccos_\star x))}$$

$$= \quad e^{\cos\left(\log e^{\mathrm{arccos}(\log x)}\right)}$$

$$= \quad e^{\cos(\mathrm{arccos}(\log x))}$$

$$= \quad e^{\log x}$$

$$= \quad x.$$

6. For $x \in \mathbb{R}_\star$, we have

$$\cot_\star\left(\mathrm{arccot}_\star x\right) \quad = \quad e^{\cot(\log(\mathrm{arccot}_\star x))}$$

$$= \quad e^{\cot\left(\log e^{\mathrm{arccot}(\log x)}\right)}$$

$$= \quad e^{\cot(\mathrm{arccot}(\log x))}$$

$$= \quad e^{\log x}$$

$$= \quad x.$$

Exercise 1.12. Prove that

$$\mathrm{arccot}_\star\left(\cot_\star x\right) = x, \quad x \in \mathbb{R}_\star.$$

Example 1.24. We will find

$$\sin_\star\left(\arctan_\star x\right), \quad x \in \mathbb{R}_\star.$$

We have

$$\sin_\star\left(\arctan_\star x\right) \quad = \quad e^{\sin(\log(\arctan_\star x))}$$

$$= \quad e^{\sin\left(\log e^{\arctan(\log x)}\right)}$$

$$= \quad e^{\sin(\arctan(\log x))}, \quad x \in \mathbb{R}_\star.$$

Let

$$y = \arctan(\log x), \quad x \in \mathbb{R}_\star.$$

Then $y \in \left(-\dfrac{\pi}{2}, \dfrac{\pi}{2}\right)$ and

$$\tan y = \log x, \quad x \in \mathbb{R}_\star$$

and

$$\sin y = (\log x) \cos y, \quad x \in \mathbb{R}_\star.$$

Hence,

$$
\begin{aligned}
1 &= \sin^2 y + \cos^2 y \\
&= (\log x)^2 \cos^2 y + \cos^2 y \\
&= \left(1 + (\log x)^2\right) \cos^2 y, \quad x \in \mathbb{R}_\star,
\end{aligned}
$$

and

$$\cos^2 y = \frac{1}{1 + (\log x)^2}, \quad x \in \mathbb{R}_\star,$$

or

$$
\begin{aligned}
\cos y &= \frac{1}{\sqrt{1 + (\log x)^2}}, \\
\sin y &= \frac{\log x}{\sqrt{1 + (\log x)^2}}, \quad x \in \mathbb{R}_\star.
\end{aligned}
$$

Consequently

$$
\begin{aligned}
\sin_\star (\arctan_\star x) &= e^{\sin y} \\
&= e^{\frac{\log x}{\sqrt{1 + (\log x)^2}}}, \quad x \in \mathbb{R}_\star.
\end{aligned}
$$

Exercise 1.13. Find

1. $\cos_\star (\arcsin_\star x), x \in \mathbb{R}_\star.$
2. $\tan_\star (\arcsin_\star x), x \in \mathbb{R}_\star.$
3. $\cot_\star (\arccos_\star x), x \in \mathbb{R}_\star.$

Answer 1.8.

1. $e^{\sqrt{1 - (\log x)^2}}, x \in \left[e^{-1}, e\right].$

2. $e^{\frac{\log x}{\sqrt{1 - (\log x)^2}}}, x \in \left(e^{-1}, e\right).$

3. $e^{\frac{\log x}{\sqrt{1 - (\log x)^2}}}, x \in \left(e^{-1}, e\right).$

1.9 Multiplicative Hyperbolic Functions

In this section, we will define the multiplicative hyperbolic functions and deduct some of their properties.

Definition 1.25. For $x \in \mathbb{R}_\star$, define

1. $\sinh_\star x = e^{\sinh(\log x)}$.
2. $\cosh_\star x = e^{\cosh(\log x)}$.
3. $\tanh_\star x = e^{\tanh(\log x)}$.
4. $\coth_\star x = e^{\coth(\log x)}$.

Now, we have listed some of the properties of the multiplicative hyperbolic functions.

1. For $x \in \mathbb{R}_\star$, we have

$$\tanh_\star x = \sinh_\star x /_\star \cosh_\star x.$$

Proof We have

$$
\begin{aligned}
\sinh_\star x /_\star \cosh_\star x &= e^{\frac{\log(\sinh_\star x)}{\log(\cosh_\star x)}} \\[2ex]
&= e^{\frac{\log\left(e^{\sinh(\log x)}\right)}{\log\left(e^{\cosh(\log x)}\right)}} \\[2ex]
&= e^{\frac{\sinh(\log x)}{\cosh(\log x)}} \\[2ex]
&= e^{\tanh(\log x)} \\[2ex]
&= \tanh_\star x.
\end{aligned}
$$

This completes the proof.

2. For $x \in \mathbb{R}_\star$, we have

$$\coth_\star x = \cosh_\star x /_\star \sinh_\star x.$$

Proof We have

$$\cosh_\star x /_\star \sinh_\star x \quad = \quad e^{\frac{\log(\cosh_\star x)}{\log(\sinh_\star x)}}$$

$$= \quad e^{\frac{\log\left(e^{\cosh(\log x)}\right)}{\log\left(e^{\sinh(\log x)}\right)}}$$

$$= \quad e^{\frac{\cosh(\log x)}{\sinh(\log x)}}$$

$$= \quad e^{\coth(\log x)}$$

$$= \quad \coth_\star x.$$

This completes the proof.

3. For $x, y \in \mathbb{R}_\star$, we have

$$\sinh_\star(x +_\star y) = \sinh_\star x \cdot_\star \cosh_\star y +_\star \sinh_\star y \cdot_\star \cosh_\star x.$$

Proof We have

$$\sinh_\star(x +_\star y) \quad = \quad \sinh_\star(xy)$$

$$= \quad e^{\sinh(\log(xy))}$$

$$= \quad e^{\sinh(\log x + \log y)}$$

$$= \quad e^{\sinh(\log x)\cosh(\log y) + \sinh(\log y)\cosh(\log x)}$$

$$= \quad e^{\sinh(\log x)\cosh(\log y)} e^{\sinh(\log y)\cosh(\log x)}$$

and

$$\sinh_\star x \cdot_\star \cosh_\star y +_\star \sinh_\star y \cdot_\star \cosh_\star x$$

$$= \quad e^{\log(\sinh_\star x)\log(\cosh_\star y)} +_\star e^{\log(\sinh_\star y)\log(\cosh_\star x)}$$

$$= \quad e^{\log\left(e^{\sinh(\log x)}\right)\log\left(e^{\cosh(\log y)}\right)}$$

$$+_\star e^{\log\left(e^{\sinh(\log y)}\right)\log\left(e^{\cosh(\log x)}\right)}$$

$$= \quad e^{\sinh(\log x)\cosh(\log y)} +_\star e^{\sinh(\log y)\cosh(\log x)}$$

$$= \quad e^{\sinh(\log x)\cosh(\log y)} e^{\sinh(\log y)\cosh(\log x)}$$

$$= \quad e^{\sinh(\log x)\cosh(\log y)+\sinh(\log y)\cosh(\log x)}.$$

Consequently

$$\sinh_\star(x +_\star y) = \sinh_\star x \cdot_\star \cosh_\star y +_\star \sinh_\star y \cdot_\star \cosh_\star x.$$

This completes the proof.

4. For $x, y \in \mathbb{R}_\star$, we have

$$\cosh_\star(x +_\star y) = \cosh_\star x \cdot_\star \cosh_\star y +_\star \sinh_\star y \cdot_\star \sinh_\star x.$$

Proof We have

$$\begin{aligned}
\cosh_\star(x +_\star y) &= \quad \cosh_\star(xy) \\
&= \quad e^{\cosh(\log(xy))} \\
&= \quad e^{\cosh(\log x + \log y)} \\
&= \quad e^{\cosh(\log x)\cosh(\log y)+\sinh(\log y)\sinh(\log x)} \\
&= \quad e^{\cosh(\log x)\cosh(\log y)} e^{\sinh(\log y)\sinh(\log x)}
\end{aligned}$$

and

$$\cosh_\star x \cdot_\star \cosh_\star y +_\star \sinh_\star y \cdot_\star \sinh_\star x$$

$$= \quad e^{\log(\cosh_\star x)\log(\cosh_\star y)} +_\star e^{\log(\sinh_\star y)\log(\sinh_\star x)}$$

$$= \quad e^{\log\left(e^{\cosh(\log x)}\right)\log\left(e^{\cosh(\log y)}\right)}$$

$$+_\star e^{\log\left(e^{\sinh(\log y)}\right)\log\left(e^{\sinh(\log x)}\right)}$$

$$= \quad e^{\cosh(\log x)\cosh(\log y)} +_\star e^{\sinh(\log y)\sinh(\log x)}$$

$$= \quad e^{\cosh(\log x)\cosh(\log y)} e^{\sinh(\log y)\sinh(\log x)}$$

$$= \quad e^{\cosh(\log x)\cosh(\log y)+\sinh(\log y)\sinh(\log x)}.$$

Consequently

$$\cosh_\star(x +_\star y) = \cosh_\star x \cdot_\star \cosh_\star y +_\star \sinh_\star y \cdot_\star \sinh_\star x.$$

This completes the proof.

1.10   Multiplicative Inverse Hyperbolic Functions

In this section, we will define the multiplicative inverse hyperbolic functions and deduct some of their properties.

Definition 1.26. For $x \in \mathbb{R}_\star$, define

1. $\sinh_\star^{-1\star} x = e^{\sinh^{-1}(\log x)}$.

2. $\cosh_\star^{-1\star} x = e^{\cosh^{-1}(\log x)}$.

3. $\tanh_\star^{-1\star} x = e^{\tanh^{-1}(\log x)}$.

4. $\coth_\star^{-1\star} x = e^{\coth^{-1}(\log x)}$.

Now, we have listed some of the properties of the multiplicative inverse hyperbolic functions.

1. $\sinh_\star \left(\sinh_\star^{-1\star} x \right) = x, x \in \mathbb{R}_\star$.

   Proof We have

$$
\begin{aligned}
\sinh_\star \left(\sinh_\star^{-1\star} x \right) &= e^{\sinh\left(\log\left(\sinh_\star^{-1\star} x \right) \right)} \\
&= e^{\sinh\left(\log e^{\sinh^{-1}(\log x)} \right)} \\
&= e^{\sinh\left(\sinh^{-1}(\log x) \right)} \\
&= e^{\log x} \\
&= x, \quad x \in \mathbb{R}_\star.
\end{aligned}
$$

   This completes the proof.

2. $\cosh_\star \left(\cosh_\star^{-1\star} x \right) = x, x \in \mathbb{R}_\star$.

   Proof We have

$$
\cosh_\star \left(\cosh_\star^{-1\star} x \right) = e^{\cosh\left(\log\left(\cosh_\star^{-1\star} x \right) \right)}
$$

$$= e^{\cosh\left(\log e^{\cosh^{-1}(\log x)}\right)}$$

$$= e^{\cosh\left(\cosh^{-1}(\log x)\right)}$$

$$= e^{\log x}$$

$$= x, \quad x \in \mathbb{R}_\star.$$

This completes the proof.

3. $\tanh_\star\left(\tanh_\star^{-1} x\right) = x, x \in \mathbb{R}_\star.$

   Proof We have

$$\tanh_\star\left(\tanh_\star^{-1} x\right) = e^{\tanh\left(\log\left(\tanh_\star^{-1} x\right)\right)}$$

$$= e^{\tanh\left(\log e^{\tanh^{-1}(\log x)}\right)}$$

$$= e^{\tanh\left(\tanh^{-1}(\log x)\right)}$$

$$= e^{\log x}$$

$$= x, \quad x \in \mathbb{R}_\star.$$

This completes the proof.

4. $\coth_\star\left(\coth_\star^{-1} x\right) = x, x \in \mathbb{R}_\star.$

   Proof We have

$$\coth_\star\left(\coth_\star^{-1} x\right) = e^{\coth\left(\log\left(\coth_\star^{-1} x\right)\right)}$$

$$= e^{\coth\left(\log e^{\coth^{-1}(\log x)}\right)}$$

$$= e^{\coth\left(\coth^{-1}(\log x)\right)}$$

$$= e^{\log x}$$

$$= x, x \in \mathbb{R}_\star.$$

This completes the proof.

1.11　Multiplicative Matrices

Definition 1.27. A matrix whose entries are elements of $\mathbb{R}_\star$ will be called a multiplicative matrix. The set of all multiplicative matrices $m \times n$ will be denoted by $\mathcal{M}_{\star m \times n}$.

Definition 1.28. Let $A = (a_{ij}) \in \mathcal{M}_{\star m \times n}$ and $B = (b_{ij}) \in \mathcal{M}_{\star m \times n}$ be multiplicative matrices and $\lambda \in \mathbb{R}_\star$. Then, we define

$$A \pm_\star B = (a_{ij} \pm_\star b_{ij}) \quad \text{and} \quad \lambda \cdot_\star A = (\lambda \cdot_\star a_{ij}).$$

Example 1.25. Let

$$A = \begin{pmatrix} 2 & \dfrac{1}{3} \\ 4 & \dfrac{1}{8} \end{pmatrix} \quad \text{and} \quad \lambda = 3.$$

Then

$$3 \cdot_\star 2 = e^{\log 3 \log 2},$$

$$3 \cdot_\star \frac{1}{3} = e^{\log 3 \log \frac{1}{3}}$$

$$= e^{-(\log 3)^2},$$

$$3 \cdot_\star 4 = e^{\log 3 \log 4}$$

$$= e^{2 \log 3 \log 2},$$

$$3 \cdot_\star \frac{1}{8} = e^{\log 3 \log \frac{1}{8}}$$

$$= e^{-3 \log 3 \log 2}.$$

Hence,

$$\lambda \cdot_\star A = \begin{pmatrix} 3 \cdot_\star 2 & 3 \cdot_\star \dfrac{1}{3} \\ 3 \cdot_\star 4 & 3 \cdot_\star \dfrac{1}{8} \end{pmatrix}$$

$$= \begin{pmatrix} e^{\log 3 \log 2} & e^{-(\log 3)^2} \\ e^{2 \log 3 \log 2} & e^{-3 \log 3 \log 2} \end{pmatrix}.$$

Example 1.26. Let

$$A = \begin{pmatrix} 1 & e \\ 2 & 3 \end{pmatrix} \quad \text{and} \quad B = \begin{pmatrix} 2 & 3 \\ \dfrac{1}{2} & \dfrac{1}{3} \end{pmatrix}.$$

Then

$$\begin{aligned} 1 +_\star 2 &= 1 \cdot 2 \\ &= 2, \\ e +_\star 3 &= e \cdot 3 \\ &= 3e, \\ 2 +_\star \frac{1}{2} &= 2 \cdot \frac{1}{2} \\ &= 1, \\ 3 +_\star \frac{1}{3} &= 3 \cdot \frac{1}{3} \\ &= 1. \end{aligned}$$

Hence,

$$\begin{aligned} A +_\star B &= \begin{pmatrix} 1 +_\star 2 & e +_\star 3 \\ 2 +_\star \dfrac{1}{2} & 3 +_\star \dfrac{1}{3} \end{pmatrix} \\ &= \begin{pmatrix} 2 & 3e \\ 1 & 1 \end{pmatrix}. \end{aligned}$$

Exercise 1.14. Let

$$A = \begin{pmatrix} \dfrac{1}{3} & \dfrac{1}{2} & 4 \\ 2 & 1 & e \\ 3 & 4 & 5 \end{pmatrix} \quad \text{and} \quad B = \begin{pmatrix} \dfrac{1}{3} & \dfrac{1}{5} & 6 \\ 2 & 3 & 1 \\ e & 2 & 4 \end{pmatrix}.$$

Find

1. $A +_\star B$.
2. $A -_\star B$.
3. $(3 \cdot_\star A) +_\star \left(\dfrac{1}{2} \cdot_\star B \right)$.

Answer 1.9.

1. $\begin{pmatrix} \dfrac{1}{9} & \dfrac{1}{10} & 24 \\ 4 & 3 & e \\ 3e & 8 & 20 \end{pmatrix}.$

2. $\begin{pmatrix} 1 & \dfrac{5}{2} & \dfrac{2}{3} \\ 1 & \dfrac{1}{3} & e \\ \dfrac{3}{e} & 2 & \dfrac{5}{4} \end{pmatrix}.$

3. $\begin{pmatrix} e^{\log 3 \log \frac{2}{3}} & e^{\log 2 \log \frac{5}{3}} & e^{\log 2 \log \frac{3}{2}} \\ e^{\log 2 \log \frac{3}{2}} & e^{-\log 2 \log 3} & 3 \\ e^{(\log 3)^2 - \log 2} & e^{\log 2 \log \frac{9}{2}} & e^{\log 3 \log 5 - 2(\log 2)^2} \end{pmatrix}.$

Definition 1.29. The matrix $I_\star \in \mathcal{M}_{n \times n}$, given by

$$I_\star = \begin{pmatrix} 1_\star & 0_\star & \cdots & 0_\star \\ 0_\star & 1_\star & \cdots & 0_\star \\ \vdots & \vdots & \vdots & \vdots \\ 0_\star & 0_\star & \cdots & 1_\star \end{pmatrix},$$

will be called the multiplicative unit matrix.

We have

$$I_\star = \begin{pmatrix} e & 1 & \cdots & 1 \\ 1 & e & \cdots & 1 \\ \vdots & \vdots & \vdots & \vdots \\ 1 & 1 & \cdots & e \end{pmatrix}.$$

Definition 1.30. The matrix $O_\star \in \mathcal{M}_{\star n \times n}$, defined by

$$O_\star = \begin{pmatrix} 0_\star & 0_\star & \cdots & 0_\star \\ 0_\star & 0_\star & \cdots & 0_\star \\ \vdots & \vdots & \vdots & \vdots \\ 0_\star & 0_\star & \cdots & 0_\star \end{pmatrix},$$

will be called the multiplicative zero matrix.

We have

$$O_\star = \begin{pmatrix} 1 & 1 & \cdots & 1 \\ 1 & 1 & \cdots & 1 \\ \vdots & \vdots & \vdots & \vdots \\ 1 & 1 & \cdots & 1 \end{pmatrix}.$$

If $A = (a_{ij}) \in \mathcal{M}_{\star m \times n}$, then

$$
\begin{aligned}
A -_\star A &= (a_{ij}) -_\star (a_{ij}) \\[2mm]
&= \left(\frac{a_{ij}}{a_{ij}} \right) \\[2mm]
&= (1) \\[2mm]
&= (0_\star) \\[2mm]
&= O_\star.
\end{aligned}
$$

Definition 1.31. Let $A \in \mathcal{M}_{\star r \times n}$, $A = (a_{ij})$, $B \in \mathcal{M}_{\star n \times m}$, $B = (b_{ij})$. Then, we define $A \cdot_\star B \in \mathcal{M}_{\star r \times m}$ as follows

$$
A \cdot_\star B = (a_{k1} \cdot_\star b_{1l} +_\star a_{k2} \cdot_\star b_{2l} +_\star \cdots +_\star a_{kn} \cdot_\star b_{nl}).
$$

Example 1.27. Let

$$
A = \begin{pmatrix} 1 & \dfrac{1}{2} & 3 \\ 2 & e & 1 \end{pmatrix}, \quad B = \begin{pmatrix} 1 & 2 \\ e & 3 \\ \dfrac{1}{4} & 5 \end{pmatrix}.
$$

We will find $A \cdot_\star B$. We have

$$
\begin{aligned}
& 1 \cdot_\star 1 +_\star \frac{1}{2} \cdot_\star e +_\star 3 \cdot_\star \frac{1}{4} \\[2mm]
&= e^{\log 1 \log 1} +_\star e^{\log \frac{1}{2} \log e} +_\star e^{\log 3 \log \frac{1}{4}} \\[2mm]
&= 1 +_\star e^{-\log 2} +_\star e^{-2 \log 3 \log 2} \\[2mm]
&= 1 \cdot e^{-\log 2} \cdot e^{-2 \log 3 \log 2} \\[2mm]
&= \frac{1}{2} e^{-2 \log 3 \log 2}
\end{aligned}
$$

and

$$
1 \cdot_\star 2 +_\star \frac{1}{2} \cdot_\star 3 +_\star 3 \cdot_\star 5
$$

$$= \quad e^{\log 1 \log 2} +_\star e^{\log \frac{1}{2} \log 3} +_\star e^{\log 3 \log 5}$$

$$= \quad 1 +_\star e^{-\log 2 \log 3} +_\star e^{\log 3 \log 5}$$

$$= \quad 1 \cdot e^{-\log 2 \log 3} \cdot e^{\log 3 \log 5}$$

$$= \quad e^{\log 3 (\log 5 - \log 2)}$$

$$= \quad e^{\log 3 \log \frac{5}{2}},$$

and

$$2 \cdot_\star 1 +_\star e \cdot_\star e +_\star 1 \cdot_\star \frac{1}{4}$$

$$= \quad e^{\log 2 \log 1} +_\star e^{\log e \log e} +_\star e^{\log 1 \log \frac{1}{4}}$$

$$= \quad 1 +_\star e +_\star 1$$

$$= \quad e,$$

and

$$2 \cdot_\star 2 +_\star e \cdot_\star 3 +_\star 1 \cdot_\star 5$$

$$= \quad e^{(\log 2)^2} +_\star e^{\log e \log 3} +_\star e^{\log 1 \log 5}$$

$$= \quad e^{(\log 2)^2} +_\star e^{\log 3}$$

$$= \quad 3 e^{(\log 2)^2}.$$

Hence,

$$A \cdot_\star B = \begin{pmatrix} \dfrac{1}{2} e^{-2 \log 3 \log 2} & e^{\log 3 \log \frac{5}{2}} \\ e & 3 e^{(\log 2)^2} \end{pmatrix}.$$

Exercise 1.15. Find $A \cdot_\star B$, where

$$A = \begin{pmatrix} 2 & 3 \\ 4 & 5 \\ 6 & 7 \end{pmatrix} \quad \text{and} \quad B = \begin{pmatrix} 2 & 1 & 3 & 6 & 7 \\ 1 & 1 & 1 & 1 & \\ \dfrac{1}{2} & \dfrac{1}{4} & \dfrac{1}{3} & \dfrac{1}{8} & 5 \end{pmatrix}.$$

Answer 1.10.

$$A \cdot_\star B = \begin{pmatrix} e^{(\log 2)^2 - \log 3 \log 2} & e^{-2\log 3 \log 2} & e^{-\log 3 \log \frac{3}{2}} & e^{-\log 2 \log \frac{9}{2}} & e^{\log 2 \log 7 + \log 3 \log 5} \\ e^{-\log 2 \log \frac{5}{4}} & e^{-2\log 2 \log 5} & e^{-\log 3 \log \frac{5}{4}} & e^{-\log 2 \log \frac{125}{36}} & e^{2\log 2 \log 7 + (\log 5)^2} \\ e^{-\log 2 \log \frac{7}{6}} & e^{-2\log 2 \log 7} & e^{-\log 3 \log \frac{7}{6}} & e^{(\log 6)^2 - 3\log 2 \log 7} & e^{\log 7 \log 30} \end{pmatrix}.$$

Definition 1.32. Let

$$A = \begin{pmatrix} a_{11} & a_{12} \\ a_{21} & a_{22} \end{pmatrix} \in \mathcal{M}_{\star 2 \times 2}.$$

We define the multiplicative determinant of A as follows

$$\det{}_\star A = a_{11} \cdot_\star a_{22} -_\star a_{21} \cdot_\star a_{12}.$$

We have

$$\begin{aligned} \det{}_\star A &= a_{11} \cdot_\star a_{22} -_\star a_{21} \cdot_\star a_{12} \\[1em] &= e^{\log a_{11} \log a_{22}} -_\star e^{\log a_{12} \log a_{21}} \\[1em] &= \frac{e^{\log a_{11} \log a_{22}}}{e^{\log a_{21} \log a_{12}}} \\[1em] &= e^{\log a_{11} \log a_{22} - \log a_{12} \log a_{21}}. \end{aligned}$$

Example 1.28. Let

$$A = \begin{pmatrix} 2 & e \\ 1 & 3 \end{pmatrix}.$$

We will find $\det_\star A$. We have

$$\begin{aligned} \det{}_\star A &= e^{\log 2 \log 3 - \log 1 \log 2} \\[1em] &= e^{\log 2 \log 3}. \end{aligned}$$

Exercise 1.16. Find $\det_\star A$, where

$$A = \begin{pmatrix} \dfrac{1}{2} & 3 \\ 4 & \dfrac{1}{5} \end{pmatrix}.$$

Answer 1.11. $e^{-\log 2 \log \frac{9}{5}}$.

Definition 1.33. Let

$$A = \begin{pmatrix} a_{11} & a_{12} \\ a_{21} & a_{22} \end{pmatrix} \in \mathcal{M}_{\star 2 \times 2}$$

and

$$\det{}_\star A = e^{\log a_{11} \log a_{22} - \log a_{12} \log a_{21}}$$

$$\neq 0_\star.$$

Then, we define the multiplicative inverse matrix $A^{-1\star}$ of the matrix A as follows

$$A^{-1\star} = (1_\star/_\star \det{}_\star A) \cdot_\star \begin{pmatrix} a_{22} & \dfrac{1}{a_{12}} \\ \dfrac{1}{a_{21}} & a_{11} \end{pmatrix}.$$

We have

$$A^{-1\star} = (e/_\star \det{}_\star A) \cdot_\star \begin{pmatrix} a_{22} & \dfrac{1}{a_{12}} \\ \dfrac{1}{a_{21}} & a_{11} \end{pmatrix}$$

$$= e^{\frac{1}{\log a_{11} \log a_{22} - \log a_{12} \log a_{21}}} \cdot_\star \begin{pmatrix} a_{22} & \dfrac{1}{a_{12}} \\ \dfrac{1}{a_{21}} & a_{11} \end{pmatrix}$$

$$= \begin{pmatrix} e^{\frac{\log a_{22}}{\log a_{11} \log a_{22} - \log a_{12} \log a_{21}}} & e^{-\frac{\log a_{12}}{\log a_{11} \log a_{22} - \log a_{12} \log a_{21}}} \\ e^{-\frac{\log a_{21}}{\log a_{11} \log a_{22} - \log a_{12} \log a_{21}}} & e^{\frac{\log a_{11}}{\log a_{11} \log a_{22} - \log a_{12} \log a_{21}}} \end{pmatrix}$$

Hence,

$$A^{-1\star} \cdot_\star A = A \cdot_\star A^{-1\star}$$

$$= \begin{pmatrix} e^{\frac{\log a_{22}}{\log a_{11} \log a_{22} - \log a_{12} \log a_{21}}} & e^{-\frac{\log a_{12}}{\log a_{11} \log a_{22} - \log a_{12} \log a_{21}}} \\ e^{-\frac{\log a_{21}}{\log a_{11} \log a_{22} - \log a_{12} \log a_{21}}} & e^{\frac{\log a_{11}}{\log a_{11} \log a_{22} - \log a_{12} \log a_{21}}} \end{pmatrix} \cdot_\star \begin{pmatrix} a_{11} & a_{12} \\ a_{21} & a_{22} \end{pmatrix}$$

$$= \begin{pmatrix} e^{\frac{\log a_{11} \log a_{22} - \log a_{12} \log a_{21}}{\log a_{11} \log a_{22} - \log a_{12} \log a_{21}}} & e^{-\frac{\log a_{22} \log a_{12} - \log a_{22} \log a_{12}}{\log a_{11} \log a_{22} - \log a_{12} \log a_{21}}} \\ e^{\frac{\log a_{21} \log a_{11} - \log a_{21} \log a_{11}}{\log a_{11} \log a_{22} - \log a_{12} \log a_{21}}} & e^{\frac{\log a_{11} \log a_{22} - \log a_{12} \log a_{21}}{\log a_{11} \log a_{22} - \log a_{12} \log a_{21}}} \end{pmatrix}$$

$$= \begin{pmatrix} e & 1 \\ 1 & e \end{pmatrix}$$

$$= \begin{pmatrix} 1_\star & 0_\star \\ 0_\star & 1_\star \end{pmatrix}$$

$$= I_\star.$$

Example 1.29. Let

$$A = \begin{pmatrix} 2 & 3 \\ 3 & 4 \end{pmatrix}.$$

We will find $A^{-1\star}$. We have

$$A^{-1\star} = \begin{pmatrix} e^{\frac{\log 4}{\log 2\log 4 - (\log 3)^2}} & e^{-\frac{\log 3}{\log 2\log 4 - (\log 3)^2}} \\ e^{-\frac{\log 3}{\log 2\log 4 - (\log 3)^2}} & e^{\frac{\log 2}{\log 2\log 4 - (\log 3)^2}} \end{pmatrix}$$

$$= \begin{pmatrix} e^{\frac{2\log 2}{2(\log 2)^2 - (\log 3)^2}} & e^{-\frac{\log 3}{2(\log 2)^2 - (\log 3)^2}} \\ e^{-\frac{\log 3}{2(\log 2)^2 - (\log 3)^2}} & e^{\frac{\log 2}{2(\log 2)^2 - (\log 3)^2}} \end{pmatrix}.$$

Exercise 1.17. Find $A^{-1\star}$, where

$$A = \begin{pmatrix} \dfrac{1}{2} & 4 \\ \dfrac{1}{3} & 9 \end{pmatrix}.$$

Answer 1.12. $A^{-1\star}$ *does not exist because* $\det_\star A = 0_\star$.

Definition 1.34. Let

$$A = \begin{pmatrix} a_{11} & a_{12} & a_{13} \\ a_{21} & a_{22} & a_{23} \\ a_{31} & a_{32} & a_{33} \end{pmatrix} \in \mathcal{M}_{\star 3\times 3}.$$

Define the multiplicative determinant of A as follows

$$\det_\star A = a_{11} \cdot_\star a_{22} \cdot_\star a_{33} +_\star a_{31} \cdot_\star a_{12} \cdot_\star a_{23}$$

$$+_\star a_{21} \cdot_\star a_{32} \cdot_\star a_{13} -_\star a_{31} \cdot_\star a_{22} \cdot_\star a_{13}$$

$$-_\star a_{11} \cdot_\star a_{32} \cdot_\star a_{23} -_\star a_{21} \cdot_\star a_{12} \cdot_\star a_{33}.$$

We have

$$\det_\star A = e^{\log a_{11} \log a_{22} \log a_{33}} +_\star e^{\log a_{31} \log a_{12} \log a_{23}}$$

$$+_\star e^{\log a_{21} \log a_{32} \log a_{13}} -_\star e^{\log a_{31} \log a_{22} \log a_{13}}$$

$$-_\star e^{\log a_{11} \log a_{32} \log a_{23}} -_\star e^{\log a_{21} \log a_{12} \log a_{33}}$$

$$= e^{\log a_{11} \log a_{22} \log a_{33} + \log a_{31} \log a_{12} \log a_{23} + \log a_{21} \log a_{32} \log a_{13} - \log a_{31} \log a_{22} \log a_{13} - \log a_{11} \log a_{32} \log a_{23} - \log a_{21} \log a_{12} \log a_{33}}.$$

Example 1.30. Let

$$A = \begin{pmatrix} e & e^2 & e^4 \\ e^3 & e^7 & e^{11} \\ e^{15} & e^8 & e^9 \end{pmatrix}.$$

Then

$$\det{}_{\star}A = e^{1 \cdot 7 \cdot 9 + 2 \cdot 11 \cdot 15 + 3 \cdot 8 \cdot 4 - 4 \cdot 7 \cdot 15 - 1 \cdot 11 \cdot 8 - 2 \cdot 3 \cdot 9}$$

$$= e^{63 + 330 + 96 - 420 - 88 - 54}$$

$$= e^{159 - 90 - 142}$$

$$= e^{69 - 142}$$

$$= e^{-73}.$$

Exercise 1.18. Let

$$A = \begin{pmatrix} 1 & e^3 & e^7 \\ 4 & 5 & e^8 \\ 6 & 7 & e^{-1} \end{pmatrix}.$$

Find $\det{}_{\star}A$.

Answer 1.13.

$$\det{}_{\star}A = e^{24 \log 6 + 14 \log 2 \log 7 - 7 \log 5 \log 6 + 6 \log 2}.$$

Definition 1.35. Let

$$A = \begin{pmatrix} a_{11} & a_{12} & a_{13} \\ a_{21} & a_{22} & a_{23} \\ a_{31} & a_{32} & a_{33} \end{pmatrix} \in \mathcal{M}_{\star 3 \times 3}.$$

Suppose that $\det{}_{\star}A \neq 0_{\star}$. Set

$$b_{11} = e^{\frac{\log a_{22} \log a_{33}}{\log a_{32} \log a_{23}}},$$

$$b_{12} = e^{\frac{\log a_{32} \log a_{13}}{\log a_{12} \log a_{33}}},$$

$$b_{13} = e^{\frac{\log a_{12} \log a_{23}}{\log a_{13} \log a_{22}}},$$

$$b_{21} = e^{\frac{\log a_{31} \log a_{23}}{\log a_{21} \log a_{33}}},$$

$$b_{22} = e^{\frac{\log a_{11} \log a_{33}}{\log a_{13} \log a_{31}}},$$

$$b_{23} = e^{\frac{\log a_{21} \log a_{13}}{\log a_{11} \log a_{23}}},$$

$$b_{31} = e^{\frac{\log a_{21} \log a_{32}}{\log a_{31} \log a_{22}}},$$

$$b_{32} = e^{\frac{\log a_{31} \log a_{12}}{\log a_{11} \log a_{32}}},$$

$$b_{33} = e^{\frac{\log a_{11} \log a_{22}}{\log a_{12} \log a_{21}}}.$$

The matrix

$$A^{-1\star} = (1_\star/_\star \det_\star A) \cdot_\star \begin{pmatrix} b_{11} & b_{12} & b_{13} \\ b_{21} & b_{22} & b_{23} \\ b_{31} & b_{32} & b_{33} \end{pmatrix}$$

is said to be the multiplicative inverse matrix of the matrix A.

Exercise 1.19. Let $A \in \mathcal{M}_{\star 3 \times 3}$ and $\det_\star A \neq 0_\star$. Prove that

$$A \cdot_\star A^{-1\star} = A^{-1\star} \cdot_\star A = I_\star.$$

Exercise 1.20. Let

$$A = \begin{pmatrix} 2 & e^3 & e^7 \\ 3 & 11 & 5 \\ e^2 & e & e^4 \end{pmatrix}.$$

Find $A^{-1\star}$.

Answer 1.14.

$$A^{-1\star} = e^{\frac{4\log 2 \log 11 + 6\log 5 + 7\log 3}{14\log 11 + \log 2 \log 5 + 12 \log 3}}$$

$$\begin{pmatrix} e^{\frac{4\log 11}{\log 5}} & e^{\frac{7}{12}} & e^{\frac{3\log 5}{7\log 11}} \\ e^{\frac{2\log 5}{4\log 3}} & e^{\frac{2\log 2}{7}} & e^{\frac{7\log 3}{\log 2 \log 5}} \\ e^{\frac{\log 3}{2\log 11}} & e^{\frac{6}{\log 2}} & e^{\frac{\log 2 \log 11}{3\log 3}} \end{pmatrix}.$$

Definition 1.36. A matrix $A \in \mathcal{M}_{\star n \times n}$ is said to be a multiplicative orthogonal matrix if

$$A \cdot_\star A^T = I_\star.$$

1.12 Advanced Practical Problems

Problem 1.1. Find

    1. $A = (3 \cdot_\star 4) -_\star (1 +_\star 5)/_\star 2$.

    2. $A = (2 -_\star 4) \cdot_\star 7 +_\star (3 +_\star 1)/_\star 8$.

    3. $A = (1 +_\star 2)/_\star 3 +_\star (4 +_\star 2)/_\star 7$.

Answer 1.15.

    1. $e^{2\log 2 \log 3 - \frac{\log 5}{\log 2}}$.

    2. $e^{-\log 2 \log 7 + \frac{\log 3}{3\log 2}}$.

    3. $e^{\frac{\log 2}{\log 3} + \frac{3\log 2}{\log 7}}$.

Problem 1.2. Solve the equations

    1. $3 -_\star x = 1$.

    2. $4 +_\star x = 7$.

    3. $7 +_\star x = 18$.

Answer 1.16.

    1. $x = 3$.

    2. $x = \dfrac{7}{4}$.

    3. $x = \dfrac{18}{7}$.

Problem 1.3. Solve the equations

    1. $2 \cdot_\star x = 7$.

    2. $4 \cdot_\star x = 8$.

    3. $3/_\star x = 5$.

Answer 1.17.

    1. $x = e^{\frac{\log 7}{\log 2}}$.

    2. $x = e^{\frac{3}{2}}$.

    3. $x = e^{\frac{\log 3}{\log 5}}$.

Problem 1.4. Solve the equations

    1. $3 +_\star (4 \cdot_\star x) = 5$.

    2. $(2 -_\star x) \cdot_\star 7 = 1$.

    3. $(4 -_\star x) \cdot_\star 2 = 1$.

Answer 1.18.

    1. $x = e^{\frac{\log \frac{5}{3}}{2\log 2}}$.

    2. $x = 2$.

    3. $x = 4$.

Problem 1.5. Compute

    1. $A = |3 -_\star 4|_\star \cdot_\star 4$.

    2. $A = |2 +_\star 1|_\star \cdot_\star |2/_\star 3|_\star$.

    3. $A = |2 \cdot_\star 4|_\star \cdot_\star |3 -_\star 1|_\star$.

Answer 1.19.

    1. $e^{\log \frac{4}{3} \log 4}$.

    2. 3.

    3. $e^{2(\log 2)^2 \log 3}$.

Problem 1.6. Compute

    1. $A = ((7 +_\star 2) \cdot_\star 3)^{2_\star}$.

    2. $A = ((6 -_\star 3) /_\star 4)^{2_\star} +_\star 2$.

    3. $A = ((5 +_\star 1) \cdot_\star 7)^{3_\star} -_\star 2 \cdot_\star 4)^{4_\star}$.

Answer 1.20.

    1. $e^{(\log 14 \log 3)^2}$.

    2. $2e^{\frac{1}{4}}$.

    3. $e^{(\log 5 \log 7)^3 - 16(\log 2)^8}$.

Problem 1.7. Find

$$A = e^2 \cdot_\star \tan_\star e^{\frac{\pi}{4}} -_\star \sin_\star e^{\frac{\pi}{2}} +_\star 2 \cdot_\star \cos_\star e^{\frac{3\pi}{4}}.$$

Answer 1.21. $e^{1 - \frac{\sqrt{2}}{2} \log 2}$.

Problem 1.8. Find

    1. $\tan_\star(\arccos_\star x)$.

    2. $\tan_\star(3 \arccos_\star x)$, $x \in \mathbb{R}_\star$.

Answer 1.22.   *1.* $e^{\frac{\sqrt{1-(\log x)^2}}{\log x}}$, $x \in [e^{-1}, e]$.

    2. $e^{\frac{\log x}{\sqrt{1-\log x}}}$, $x \in (e^{-1}, e)$.

Problem 1.9. Let

$$A = \begin{pmatrix} 4 & 2 & 3 \\ 3 & 1 & e \\ 2 & 5 & \dfrac{1}{2} \end{pmatrix} \quad \text{and} \quad \begin{pmatrix} 2 & 3 & 4 \\ 2 & e & e \\ 3 & 1 & \dfrac{1}{4} \end{pmatrix}.$$

Find

    1. $A +_\star B$.

    2. $A -_\star B$.

    3. $A \cdot_\star B$.

Answer 1.23.

    1. $\begin{pmatrix} 8 & 6 & 12 \\ 6 & e & e^2 \\ 6 & 5 & \dfrac{1}{8} \end{pmatrix}.$

    2. $\begin{pmatrix} 2 & \dfrac{2}{3} & \dfrac{3}{4} \\ \dfrac{3}{2} & \dfrac{1}{e} & 1 \\ \dfrac{2}{3} & 5 & 2 \end{pmatrix}.$

    3. $\begin{pmatrix} e^{3(\log 2)^2+(\log 3)^2} & e^{\log 2(1+2\log 3)} & e^{4(\log 2)^2+\log 2-2\log 2\log 3} \\ e^{\log 3(1+\log 2)} & e^{(\log 3)^2} & e^{2\log 2(\log 3-1)} \\ e^{(\log 2)^2+\log 2\log \frac{5}{3}} & e^{\log 2\log 3+\log 5} & e^{4(\log 2)^2+\log 5} \end{pmatrix}.$

Problem 1.10. Find $\det_\star A$, where

$$A = \begin{pmatrix} 4 & \dfrac{1}{8} \\ \dfrac{1}{9} & \dfrac{1}{11} \end{pmatrix}.$$

Answer 1.24. $e^{-2\log 2\log 297}$.

Problem 1.11. Let

$$A = \begin{pmatrix} 4 & e^{-2} & 7 \\ 1 & 3 & 8 \\ e^7 & e^3 & e \end{pmatrix}.$$

Find $\det_\star A$.

Answer 1.25. $e^{2\log 2\log 3-42\log 2-7\log 3\log 7-18(\log 2)^2}$.

Problem 1.12. Find $A^{-1\star}$, where

$$A = \begin{pmatrix} 3 & \dfrac{1}{5} \\ 4 & e \end{pmatrix}.$$

Answer 1.26. $A^{-1\star} = \begin{pmatrix} e^{1+\frac{1}{\log 3 + 2\log 2\log 5}} & 5e^{\frac{1}{\log 3 + 2\log 2\log 5}} \\ \dfrac{1}{4}e^{\frac{1}{\log 3 + 2\log 2\log 5}} & 3e^{\frac{1}{\log 3 + 2\log 2\log 5}} \end{pmatrix}.$

2

Multiplicative Differentiation

In this chapter, we introduce the multiplicative derivative of a real-valued function. We deduct some of its properties such as multiplicative differentiation of multiplicative sum of two real-valued functions, multiplicative differentiation of multiplicative product and multiplicative quotient of two real-valued functions. In this chapter, we define multiplicative differentials. They are given some criteria for monotonicity of a real-valued function and local extremum of a real-valued function. In this chapter, we also deduce the multiplicative Rolle theorem, Lagrange theorem and Cauchy theorem as well as the multiplicative Taylor formula.

2.1 Definition

Let $A \subseteq \mathbb{R}_\star$ and $f \in \mathscr{C}^1(A)$.

Definition 2.1. We define the first multiplicative derivative of f at $x \in A$, which will be denoted by $f^\star(x)$, as follows

$$f^\star(x) = \lim_{h \to 0_\star} (f(x +_\star h) -_\star f(x))/_\star h.$$

We have

$$
\begin{aligned}
f^\star(x) &= \lim_{h \to 0_\star} (f(x +_\star h) -_\star f(x))/_\star h \\[2mm]
&= \lim_{h \to 1} (f(xh) -_\star f(x))/_\star h \\[2mm]
&= \lim_{h \to 1} \left(\frac{f(xh)}{f(x)} \right)/_\star h \\[2mm]
&= \lim_{h \to 1} e^{\frac{\log \frac{f(xh)}{f(x)}}{\log h}} \\[2mm]
&= \lim_{h \to 1} e^{\frac{xhf(x)f'(xh)}{f(xh)f(x)}}
\end{aligned}
$$

DOI: 10.1201/9781003299080-2

$$= e^{\frac{xf'(x)}{f(x)}}, \quad x \in A.$$

Example 2.1. Let

$$f(x) = 1 + x + x^2, \quad x \in \mathbb{R}_\star.$$

We have

$$f'(x) = 1 + 2x, \quad x \in \mathbb{R}_\star,$$

and

$$f^\star(x) = e^{\frac{x(1+2x)}{1+x+x^2}}, \quad x \in \mathbb{R}_\star.$$

Example 2.2. Let

$$f(x) = \frac{1-x}{1+x^2}, \quad x \in (0,1).$$

Then

$$
\begin{aligned}
f'(x) &= \frac{-(1+x^2) - 2x(1-x)}{(1+x^2)^2} \\[2mm]
&= -\frac{1+x^2+2x-2x^2}{(1+x^2)^2} \\[2mm]
&= -\frac{1+2x-x^2}{(1+x^2)^2} \\[2mm]
&= \frac{x^2-2x-1}{(1+x^2)^2}, \quad x \in (0,1).
\end{aligned}
$$

Hence,

$$
\begin{aligned}
f^\star(x) &= e^{\left(\frac{1-x}{1+x^2}\right)\frac{x(x^2-2x-1)}{(1+x^2)^2}} \\[3mm]
&= e^{\frac{x(x^2-2x-1)}{(1-x)(1+x^2)}}, \quad x \in (0,1).
\end{aligned}
$$

Example 2.3. Let

$$f(x) = \arcsin x + x^2 + 10, \quad x \in \left(\frac{1}{3}, \frac{1}{2}\right).$$

Then

$$f'(x) = \frac{1}{\sqrt{1-x^2}} + 2x, \quad x \in \left(\frac{1}{3}, \frac{1}{2}\right),$$

and

$$f^\star(x) = e^{\frac{x\left(\frac{1}{\sqrt{1-x^2}} + 2x\right)}{\arcsin x + x^2 + 10}}, \quad x \in \left(\frac{1}{3}, \frac{1}{2}\right).$$

Exercise 2.1. Let

1. $f(x) = 1 + x^4, x \in \mathbb{R}_\star$.
2. $f(x) = \dfrac{1+x}{1+2x}, x \in \mathbb{R}_\star$.
3. $f(x) = (\cos x)^2 + 10, x \in \mathbb{R}_\star$.

Find $f^\star(x), x \in \mathbb{R}_\star$.

Answer 2.1.   *1.* $f_\star^\star(x) = e^{\frac{4x^4}{1+x^4}}, x \in \mathbb{R}_\star$.

2. $f_\star^\star(x) = e^{-\frac{x}{(1+x)(1+2x)}}, x \in \mathbb{R}_\star$.

3. $f_\star^\star(x) = e^{-\frac{2x\sin(2x)}{(\cos x)^2 + 10}}, x \in \mathbb{R}_\star$.

Now, we will find $f_\star^\star(x), x \in A$. We have

$$f_\star(x) = e^{f(\log x)}, \quad x \in A.$$

Then

$$f_\star^\star(x) \;=\; e^{\frac{xe^{f(\log x)} f'(\log x) \frac{1}{x}}{e^{f(\log x)}}}$$

$$=\; e^{f'(\log x)}$$

$$=\; \left(f'\right)_\star(x), \quad x \in A.$$

Example 2.4. Let
$$f(x) = \cos x, \quad x \in \mathbb{R}_\star.$$

Then
$$f'(x) = -\sin x, \quad x \in \mathbb{R}_\star.$$

Hence,
$$f_\star^\star(x) = e^{-\sin(\log x)}, \quad x \in \mathbb{R}_\star.$$

Example 2.5. Let
$$f(x) = 1 + x + x^2, \quad x \in \mathbb{R}_\star.$$

Then
$$f'(x) = 1 + 2x, \quad x \in \mathbb{R}_\star,$$

and

$$f_\star^\star(x) \;=\; e^{f'(\log x)}$$

$$=\; e^{1 + 2\log x}$$

$$=\; ex^2, \quad x \in \mathbb{R}_\star.$$

Example 2.6. Let

$$f(x) = \frac{1+2x}{1+x}, \quad x \in \mathbb{R}_\star.$$

Then

$$f'(x) \quad = \quad \frac{2(1+x) - (1+2x)}{\cdot (1+x)^2}$$

$$= \quad \frac{1}{(1+x)^2}, \quad x \in \mathbb{R}_\star.$$

Hence,

$$f_\star^\star(x) \quad = \quad e^{f'(\log x)}$$

$$= \quad e^{\frac{1}{(1+\log x)^2}}, \quad x \in \mathbb{R}_\star.$$

Exercise 2.2. Let

  1. $f(x) = \sin(2x) + \cos x, x \in \mathbb{R}_\star$.
  2. $f(x) = x^3, x \in \mathbb{R}_\star$.
  3. $f(x) = x^k, x \in \mathbb{R}_\star$.

Find $f_\star^\star(x), x \in \mathbb{R}_\star$.

Answer 2.2.   *1.* $f_\star^\star(x) = e^{2\cos(2\log x) - \sin(\log x)}, x \in \mathbb{R}_\star$.

  2. $f_\star^\star(x) = e^{3(\log x)^2}, x \in \mathbb{R}_\star$.

  3. $f_\star^\star(x) = e^{k(\log x)^{k-1}}, x \in \mathbb{R}_\star$.

Below, we list the first multiplicative derivatives of the basic elementary functions.

  1. $c^\star = 1, c = \text{const}$.
  2. $(x^k)^\star = e^k, x \in \mathbb{R}_\star$.
  3. $(a^x)^\star = a^x, x \in \mathbb{R}_\star$.
  4. $(e^x)^\star = e^x, x \in \mathbb{R}_\star$.
  5. $\log^\star x = e^{\frac{1}{\log x}}, x \in \mathbb{R}_\star$.
  6. $\sin^\star x = e^{x \cot x}, x \in \mathbb{R}_\star$.
  7. $\cos^\star x = e^{-x \tan x}, x \in \mathbb{R}_\star$.
  8. $\tan^\star x = e^{\frac{2x}{\sin(2x)}}, x \in \mathbb{R}_\star$.
  9. $\cot^\star x = e^{-\frac{2x}{\sin(2x)}}, x \in \mathbb{R}_\star$.
  10. $\arcsin^\star x = e^{\frac{x}{\sqrt{1-x^2}\arcsin x}}, x \in \mathbb{R}_\star$.

11. $\arccos^{\star} x = e^{-\frac{x}{\sqrt{1-x^2}\arccos x}}$, $x \in \mathbb{R}_{\star}$.

12. $\arctan^{\star} x = e^{\frac{x}{(1+x^2)\arctan x}}$, $x \in \mathbb{R}_{\star}$.

13. $\operatorname{arccot}^{\star} x = e^{-\frac{x}{(1+x^2)\operatorname{arccot} x}}$, $x \in \mathbb{R}_{\star}$.

The space of all functions $f : A \to \mathbb{R}$ that are continuous on A and have continuous first multiplicative derivatives on A is denoted by $\mathscr{C}^1_{\star}(A)$.

2.2 Properties

In this section, we will deduct some of the properties of the first multiplicative derivative. Let $f, g \in \mathscr{C}^1_{\star}(A)$ and $a, b \in \mathbb{R}_{\star}$.

1. $(a \cdot_{\star} f)^{\star}(x) = a \cdot_{\star} f^{\star}(x)$, $x \in A$.

   Proof We have

   $$
   \begin{aligned}
   (a \cdot_{\star} f)(x) &= a \cdot_{\star} f(x) \\[2mm]
   &= e^{\log a \log f(x)}, \quad x \in A,
   \end{aligned}
   $$

   and

   $$
   \begin{aligned}
   (a \cdot_{\star} f)^{\star}(x) &= e^{x \log a \frac{f'(x) e^{\log a \log f(x)}}{f(x) e^{\log a \log f(x)}}} \\[2mm]
   &= e^{x \log a \frac{f'(x)}{f(x)}} \\[2mm]
   &= a \cdot_{\star} e^{x \frac{f'(x)}{f(x)}} \\[2mm]
   &= a \cdot_{\star} f^{\star}(x), \quad x \in A.
   \end{aligned}
   $$

   This completes the proof.

2. $(af)^{\star}(x) = f^{\star}(x)$, $x \in A$.

   Proof We have

   $$
   \begin{aligned}
   (af)^{\star}(x) &= e^{ax \frac{f'(x)}{af(x)}} \\[2mm]
   &= e^{x \frac{f'(x)}{f(x)}}
   \end{aligned}
   $$

$$= f^\star(x), \quad x \in A.$$

This completes the proof.

3. $(f +_\star g)^\star(x) = f^\star(x) +_\star g^\star(x), x \in A.$

   Proof We have

   $$(f +_\star g)(x) = f(x) +_\star g(x)$$

   $$= f(x)g(x), \quad x \in A,$$

   and

   $$(f +_\star g)^\star(x) = e^{x\frac{(fg)'(x)}{f(x)g(x)}}$$

   $$= e^{x\frac{f'(x)g(x)+g'(x)f(x)}{f(x)g(x)}}$$

   $$= e^{x\frac{f'(x)}{f(x)}+x\frac{g'(x)}{g(x)}}$$

   $$= e^{x\frac{f'(x)}{f(x)}} +_\star e^{x\frac{g'(x)}{g(x)}}$$

   $$= f^\star(x) +_\star g^\star(x), \quad x \in A.$$

   This completes the proof.

4. $(f -_\star g)^\star(x) = f^\star(x) -_\star g^\star(x), x \in A.$

   Proof We have

   $$(f -_\star g)(x) = f(x) -_\star g(x)$$

   $$= \frac{f(x)}{g(x)}, \quad x \in A,$$

   and

   $$(f -_\star g)^\star(x) = e^{x\frac{\left(\frac{f}{g}\right)'(x)}{\left(\frac{f}{g}\right)(x)}}$$

   $$= e^{x\frac{f'(x)g(x)-f(x)g'(x)}{(g(x))^2 f(x)}g(x)}$$

$$= e^{x\frac{f'(x)}{f(x)} - x\frac{g'(x)}{g(x)}}$$

$$= e^{x\frac{f'(x)}{f(x)}} -_\star e^{x\frac{g'(x)}{g(x)}}$$

$$= f^\star(x) -_\star g^\star(x), \quad x \in A.$$

This completes the proof.

5. $(f \cdot_\star g)^\star(x) = f^\star(x) \cdot_\star g(x) +_\star f(x) \cdot_\star g^\star(x), x \in A.$

   Proof We have

$$(f \cdot_\star g)(x) = f(x) \cdot_\star g(x)$$

$$= e^{\log f(x) \log g(x)}, \quad x \in A,$$

and

$$(f \cdot_\star g)^\star(x) = e^{x\frac{\left(\frac{f'(x)}{f(x)} \log g(x) + \log f(x)\frac{g'(x)}{g(x)}\right)e^{\log f(x)\log g(x)}}{e^{\log f(x)\log g(x)}}}$$

$$= e^{x\frac{f'(x)}{f(x)} \log g(x) + x\frac{g'(x)}{g(x)} \log f(x)}$$

$$= e^{x\frac{f'(x)}{f(x)} \log g(x)} +_\star e^{x\frac{g'(x)}{g(x)} \log f(x)}$$

$$= e^{x\frac{f'(x)}{f(x)}} \cdot_\star g(x) +_\star e^{x\frac{g'(x)}{g(x)}} \cdot_\star f(x)$$

$$= f^\star(x) \cdot_\star g(x) +_\star f(x) \cdot_\star g^\star(x), \quad x \in A.$$

This completes the proof.

6. $(f/_\star g)^\star(x) = (f^\star(x) \cdot_\star g(x) -_\star f(x) \cdot_\star g^\star(x))/_\star(g(x))^{2\star}, x \in A.$

   Proof We have

$$(f/_\star g)(x) = f(x)/_\star g(x)$$

$$= e^{\frac{\log f(x)}{\log g(x)}}, \quad x \in A,$$

and

$$(f/_\star g)^\star(x) \quad = \quad e^{\,x\dfrac{\left(e^{\frac{\log f(x)}{\log g(x)}}\right)'}{e^{\frac{\log f(x)}{\log g(x)}}}}$$

$$= \quad e^{\,x\dfrac{\left(\frac{\log f(x)}{\log g(x)}\right)' e^{\frac{\log f(x)}{\log g(x)}}}{e^{\frac{\log f(x)}{\log g(x)}}}}$$

$$= \quad e^{\,x\left(\frac{\log f(x)}{\log g(x)}\right)'}$$

$$= \quad e^{\,x\dfrac{\frac{f'(x)}{f(x)}\log g(x) - \frac{g'(x)}{g(x)}\log f(x)}{(\log g(x))^2}}$$

$$= \quad e^{\,x\left(\frac{f'(x)}{f(x)}\log g(x) - \frac{g'(x)}{g(x)}\log f(x)\right)}/_\star e^{(\log g(x))^2}$$

$$= \quad \left(e^{\,x\frac{f'(x)}{f(x)}}\cdot_\star g(x) -_\star e^{\,x\frac{g'(x)}{g(x)}}\cdot_\star f(x)\right)/_\star (g(x))^{2\star}$$

$$= \quad (f^\star(x)\cdot_\star g(x) -_\star f(x)\cdot_\star g^\star(x))/_\star (g(x))^{2\star}, \quad x \in A.$$

This completes the proof.

7. $(f \circ g)^\star(x) = (f^\star(g(x)))\cdot_\star g^\star(x), x \in A.$

Proof We have
$$(f \circ g)(x) = f(g(x)), \quad x \in A,$$

and

$$(f \circ g)^\star(x) \quad = \quad e^{\,x\frac{f'(g(x))g'(x)}{f(g(x))}}$$

$$= \quad e^{\,g(x)\frac{f'(g(x))}{f(g(x))}\times\frac{g'(x)}{g(x)}}$$

$$= \quad (f^\star(g(x)))\cdot_\star g^\star(x), \quad x \in A.$$

This completes the proof.

Example 2.7. We will find $(\sin(x^2+1))^\star$, $x \in \mathbb{R}_\star$. We have

$$(\sin(x^2+1))^\star \quad = \quad e^{(x^2+1)\cot(x^2+1)}\cdot_\star e^{\frac{2x^2}{x^2+1}}, \quad x \in \mathbb{R}_\star.$$

Example 2.8. We have

$$(\log(f(x)))^\star = e^{\frac{1}{\log f(x)}} \cdot_\star f^\star(x), \quad x \in A.$$

2.3   Higher Order Multiplicative Derivatives

Suppose that $A \subseteq \mathbb{R}_\star$ and $f : A \to \mathbb{R}$, $f \in \mathscr{C}^k(A)$, and $k \in \mathbb{N}$.

Definition 2.2. We define

$$f^{\star(k)}(x) = \left(f^{\star(k-1)}\right)^\star(x), \quad x \in A.$$

The space of all functions that have continuous multiplicative derivatives $f^{\star(l)}$, $l \in \{0,\ldots,k\}$, on A, will be denoted by $\mathscr{C}_\star^k(A)$.

We have

$$
\begin{aligned}
f^{\star\star}(x) &= e^{x \frac{(f^\star)'(x)}{f^\star(x)}} \\[2mm]
&= e^{x \frac{\left(e^{x \frac{f'(x)}{f(x)}}\right)'}{e^{x \frac{f'(x)}{f(x)}}}} \\[2mm]
&= e^{x \frac{\frac{(f'(x)+xf''(x))f(x)-x(f'(x))^2}{(f(x))^2} e^{x \frac{f'(x)}{f(x)}}}{e^{x \frac{f'(x)}{f(x)}}}} \\[2mm]
&= e^{\frac{x\left(f'(x)f(x)+xf''(x)f(x)-x(f'(x))^2\right)}{(f(x))^2}} \\[2mm]
&= e^{x\left(\frac{f'(x)}{f(x)}+x \frac{f''(x)f(x)-(f'(x))^2}{(f(x))^2}\right)} \\[2mm]
&= e^{x\left(\frac{f'(x)}{f(x)}+x \frac{d}{dx}\left(\frac{f'(x)}{f(x)}\right)\right)}, \quad x \in A.
\end{aligned}
$$

Example 2.9. Let

$$f(x) = 1 + x^2 + x^3, \quad x \in \mathbb{R}_\star.$$

Then

$$f'(x) = 2x + 3x^2,$$

$$f''(x) = 2 + 6x, \quad x \in \mathbb{R}_\star.$$

Hence,

$$f'(x)f(x) + xf''(x)f(x) - x(f'(x))^2 = (2x + 3x^2)(1 + x^2 + x^3) + x(2 + 6x)(1 + x^2 + x^3)$$

$$- x(2x + 3x^2)^2$$

$$= 2x + 2x^3 + 2x^4 + 3x^2 + 3x^4 + 3x^5$$

$$+ 2x + 2x^3 + 2x^4 + 6x^2 + 6x^4 + 6x^5$$

$$- 4x^3 - 12x^4 - 9x^5$$

$$= 4x + 9x^2 + x^4$$

$$= x(4 + 9x + x^3), \quad x \in \mathbb{R}_\star,$$

and

$$f^{\star\star}(x) = e^{\frac{x^2(4 + 9x + x^3)}{(1 + x^2 + x^3)^2}}, \quad x \in \mathbb{R}_\star.$$

Example 2.10. Let

$$f(x) = \sin x, \quad x \in \left(\frac{1}{3}, \frac{1}{2}\right).$$

We have

$$f'(x) = \cos x,$$

$$f''(x) = -\sin x, \quad x \in \left(\frac{1}{3}, \frac{1}{2}\right),$$

and

$$f'(x)f(x) + xf''(x)f(x) - x(f'(x))^2 = \sin x \cos x - x(\sin x)^2 - x(\cos x)^2$$

$$= \sin x \cos x - x, \quad x \in \left(\frac{1}{3}, \frac{1}{2}\right),$$

and

$$f^{\star\star}(x) = e^{\frac{x(\sin x \cos x - x)}{(\sin x)^2}}, \quad x \in \left(\frac{1}{3}, \frac{1}{2}\right).$$

Example 2.11. Let

$$f(x) = \frac{1+x}{1+2x}, \quad x \in \mathbb{R}_\star.$$

Then

$$f'(x) = \frac{1+2x-2(1+x)}{(1+2x)^2}$$

$$= -\frac{1}{(1+2x)^2},$$

$$f''(x) = \frac{4(1+2x)}{(1+2x)^4}$$

$$= \frac{4}{(1+2x)^3}, \quad x \in \mathbb{R}_\star,$$

and

$$f'(x)f(x) + xf''(x)f(x) - x(f'(x))^2 = -\frac{(1+x)}{(1+2x)^3} + \frac{4x(1+x)}{(1+2x)^4}$$

$$-\frac{x}{(1+2x)^4}$$

$$= \frac{-(1+x)(1+2x) + 4x(1+x) - x}{(1+2x)^4}$$

$$= \frac{-1-2x-x-2x^2+4x+4x^2-x}{(1+2x)^4}$$

$$= \frac{-1+2x^2}{(1+2x)^4}, \quad x \in \mathbb{R}_\star,$$

and

$$f^{\star\star}(x) = e^{\frac{x(-1+2x^2)}{(1+2x)^4} \cdot \frac{1}{\frac{(1+x)^2}{(1+2x)^2}}}$$

$$= e^{\frac{x(-1+2x^2)}{(1+x)^2(1+2x)^2}}, \quad x \in \mathbb{R}_\star.$$

Exercise 2.3. Let $f(x) = x^2, x \in \mathbb{R}_\star$. Find

$$f^\star(x), \quad f^{\star\star}(x), \quad x \in \mathbb{R}_\star.$$

Answer 2.3.   *1.* $f^\star(x) = e^2, x \in \mathbb{R}_\star$.

         2. $f^{\star\star}(x) = 1, x \in \mathbb{R}_\star$.

2.4 Multiplicative Differentials

Suppose that $A \subseteq \mathbb{R}_\star$, $f : A \to \mathbb{R}_\star$, $f \in \mathscr{C}_\star^k(A)$, $k \in \mathbb{N}$, $k \geq 2$.

Definition 2.3. We define

$$d_\star x = e^{d(\log x)}$$

$$= e^{\frac{1}{x}dx}, \quad x \in A,$$

and

$$d_\star f(x) = e^{d(\log(f(x)))}$$

$$= e^{\frac{f'(x)}{f(x)}dx}, \quad x \in A,$$

and

$$f^\star(x) = d_\star f(x)/_\star d_\star x, \quad x \in A.$$

By the above definition, it follows

$$f^\star(x) = d_\star f(x)/_\star d_\star x$$

$$= e^{\frac{f'(x)}{f(x)}dx}/_\star e^{\frac{1}{x}dx}$$

$$= e^{\frac{\log_e e^{\frac{f'(x)}{f(x)}dx}}{\log_e e^{\frac{1}{x}dx}}}$$

$$= e^{\frac{\frac{f'(x)}{f(x)}dx}{\frac{1}{x}dx}}$$

$$= e^{x\frac{f'(x)}{f(x)}dx}, \quad x \in A.$$

Moreover,

$$d_\star f(x) = f^\star(x) \cdot_\star d_\star x, \quad x \in A.$$

Definition 2.4. We define

$$d_\star^k f(x) = d_\star \left(d_\star^{k-1} f(x) \right), \quad x \in A.$$

In fact, we have

$$d_\star^2 f(x) = d_\star (d_\star f(x))$$

$$= d_\star(f(x) \cdot_\star d_\star x)$$

$$= f^{\star\star}(x) \cdot_\star d_\star x^{2\star}, \quad x \in A,$$

and so on,

$$d_\star^k f(x) = f^{\star(k)}(x) \cdot_\star d_\star x^{k\star}, \quad x \in A.$$

Example 2.12. Let

$$f(x) = x + 1, \quad x \in \mathbb{R}_\star.$$

Then

$$f'(x) = 1,$$

$$f^\star(x) = e^{\frac{x}{x+1}},$$

$$f^{\star\star}(x) = e^{\;x\frac{\left(e^{\frac{x}{x+1}}\right)'}{e^{\frac{x}{x+1}}}}$$

$$= e^{\frac{x}{(x+1)^2}}, \quad x \in \mathbb{R}_\star.$$

Hence,

$$d_\star f(x) = e^{\frac{1}{x+1}} \cdot_\star d_\star x,$$

$$d_\star^2 f(x) = e^{\frac{x}{(x+1)^2}} \cdot_\star d_\star x^{2\star}, \quad x \in \mathbb{R}_\star.$$

Now, suppose that

$$x = x(t),$$

$$y = y(t), \quad t \in [\alpha, \beta] \subset \mathbb{R}_\star,$$

and $x : [\alpha, \beta] \to A$ is a bijection. Then

$$d_\star y /_\star d_\star x = (d_\star y /_\star d_\star t) /_\star (d_\star x /_\star d_\star t)$$

$$= y^\star(t) /_\star x^\star(t), \quad t \in [\alpha, \beta].$$

Example 2.13. Let

$$x(t) = t^2 + t + 1,$$

$$y(t) = t^3 + t, \quad t \in [1,2].$$

We have

$$x^\star(t) = e^{t \frac{x'(t)}{x(t)}}$$

$$= e^{t \frac{2t+1}{t^2+t+1}}$$

$$= e^{\frac{2t^2+t}{t^2+t+1}}, \quad t \in [1,2],$$

and

$$y^\star(t) = e^{t \frac{y'(t)}{y(t)}}$$

$$= e^{t \frac{3t^2+1}{t^3+t}}$$

$$= e^{\frac{3t^2+1}{t^2+1}}, \quad t \in [1,2],$$

and

$$d_\star y / _\star dx^\star = e^{\frac{3t^2+1}{t^2+1}} / _\star e^{\frac{2t^2+t}{t^2+t+1}}$$

$$= e^{\frac{(3t^2+1)(t^2+t+1)}{(t^2+1)(2t^2+t)}}, \quad t \in [1,2].$$

Example 2.14. Let

$$x(t) = 3 + t + \sin t,$$

$$y(t) = 4 + \cos t + 7t, \quad t \in [2, 10].$$

We have

$$x^\star(t) = e^{t \frac{x'(t)}{x(t)}}$$

$$= e^{t \frac{(1+\cos t)}{3+t+\sin t}}, \quad t \in [2, 10],$$

and

$$y^\star(t) = e^{t \frac{y'(t)}{y(t)}}$$

$$= e^{t \frac{7-\sin t}{4+\cos t+7t}}, \quad t \in [2, 10].$$

Then

$$d_\star y /_\star d_\star x = e^{t \frac{7-\sin t}{4+\cos t+7t}} /_\star e^{t \frac{(1+\cos t)}{3+t+\sin t}}$$

$$= e^{\frac{(7-\sin t)(3+t+\sin t)}{(4+\cos t+7t)(1+\cos t)}}, \quad t \in [2, 10].$$

Example 2.15. Let

$$x(t) = \frac{1}{1+t},$$

$$y(t) = \frac{1}{2+t}, \quad t \in \mathbb{R}_\star.$$

We have

$$x^\star(t) = e^{t \frac{x'(t)}{x(t)}}$$

$$= e^{t \frac{-\frac{1}{(1+t)^2}}{\frac{1}{1+t}}}$$

$$= e^{-\frac{t}{1+t}}, \quad t \in \mathbb{R}_\star,$$

and

$$y^\star(t) = e^{t \frac{y'(t)}{y(t)}}$$

$$= e^{t \frac{-\frac{1}{(2+t)^2}}{\frac{1}{2+t}}}$$

$$= e^{-\frac{t}{2+t}}, \quad t \in \mathbb{R}_\star,$$

and

$$d_\star y /_\star d_\star x = e^{-\frac{t}{2+t}} /_\star e^{-\frac{t}{1+t}}$$

$$= e^{\frac{1+t}{2+t}}, \quad t \in \mathbb{R}_\star.$$

Hence,

$$d_\star^2 y /_\star d_\star x^{2\star} = (d_\star /_\star d_\star x) \left(e^{\frac{1+t}{2+t}} \right)$$

$$= \left((d_\star/_\star d_\star t) \left(e^{\frac{1+t}{2+t}} \right) \right)/_\star (d_\star x/_\star d_\star t)$$

$$= \left(e^{t\frac{\left(e^{\frac{1+t}{2+t}} \right)'}{e^{\frac{1+t}{2+t}}}} \right)/_\star \left(e^{-\frac{t}{1+t}} \right)$$

$$= e^{t\left(\frac{2+t-1-t}{(2+t)^2} \right)}/_\star \left(e^{-\frac{t}{t+1}} \right)$$

$$= e^{\frac{t}{(2+t)^2}}/_\star e^{-\frac{t}{1+t}}$$

$$= e^{\frac{\frac{t}{(2+t)^2}}{-\frac{t}{1+t}}}$$

$$= e^{-\frac{1+t}{(2+t)^2}}, \quad t \in \mathbb{R}_\star.$$

Exercise 2.4. Let

$$x(t) = \frac{1}{2+t},$$

$$y(t) = \frac{1}{3+t}, \quad t \in \mathbb{R}_\star.$$

Find

$$d^{\star 2}y/_\star d_\star x^{2\star}.$$

Answer 2.4.

$$e^{-\frac{2+t}{(3+t)^2}}, \quad t \in \mathbb{R}_\star.$$

2.5   Monotone Functions

In this section, we will give some criteria for increasing and decreasing functions. Let $A \subseteq \mathbb{R}_\star$, $f \in \mathscr{C}_\star^1(A)$.

Theorem 2.1. *Let $f(x) > 0$, $f^\star(x) \geq 0_\star$, $x \in A$. Then f is an increasing function on A.*

Proof We have

$$1 = 0_\star$$

$$\leq \; f^\star(x)$$

$$= \; e^{x \frac{f'(x)}{f(x)}}, \quad x \in A.$$

Thus,

$$x\frac{f'(x)}{f(x)} \geq 0, \quad x \in A.$$

Since $f(x) > 0$, $x \in A$, we conclude that

$$f'(x) \geq 0, \quad x \in A.$$

Hence, we obtain that f is an increasing function on A. This completes the proof.

Theorem 2.2. *Let* $f(x) < 0$, $f^\star(x) \geq 0_\star$, $x \in A$. *Then* f *is a decreasing function on* A.

    Proof We have

$$1 = 0_\star$$

$$\leq f^\star(x)$$

$$= e^{x\frac{f'(x)}{f(x)}}, \quad x \in A.$$

Thus,

$$x\frac{f'(x)}{f(x)} \geq 0, \quad x \in A.$$

Since $f(x) < 0$, $x \in A$, we conclude that

$$f'(x) \leq 0, \quad x \in A.$$

Hence, we obtain that f is a decreasing function on A. This completes the proof.

Theorem 2.3. *Let* $f(x) > 0$, $f^\star(x) \leq 0_\star$, $x \in A$. *Then* f *is a decreasing function on* A.

    Proof We have

$$1 = 0_\star$$

$$\geq f^\star(x)$$

$$= e^{x\frac{f'(x)}{f(x)}}, \quad x \in A.$$

Thus,

$$x\frac{f'(x)}{f(x)} \leq 0, \quad x \in A.$$

Since $f(x) > 0$, $x \in A$, we conclude that

$$f'(x) \leq 0, \quad x \in A.$$

Hence, we obtain that f is a decreasing function on A. This completes the proof.

Theorem 2.4. *Let $f(x) < 0$, $f^*(x) \leq 0_*$, $x \in A$. Then f is an increasing function on A.*

    Proof We have

$$1 = 0_*$$

$$\geq f^*(x)$$

$$= e^{x\frac{f'(x)}{f(x)}}, \quad x \in A.$$

Thus,

$$x\frac{f'(x)}{f(x)} \leq 0, \quad x \in A.$$

Since $f(x) < 0$, $x \in A$, we conclude that

$$f'(x) \geq 0, \quad x \in A.$$

Hence, we obtain that f is an increasing function on A. This completes the proof.

Example 2.16. Let

$$f(x) = 1 + x^2, \quad x \in \mathbb{R}_*.$$

Then

$$f'(x) = 2x,$$

$$f(x) > 0,$$

$$f^*(x) = e^{\frac{2x^2}{x^2}}$$

$$= e^2$$

$$> 0_*, \quad x \in \mathbb{R}_*.$$

Thus, f is an increasing function on $\mathbb{R}_*$.

Example 2.17. Let

$$f(x) = \frac{1+x}{1+x^2}, \quad x \in \mathbb{R}_\star.$$

Then

$$
\begin{aligned}
f'(x) &= \frac{1+x^2 - 2x(1+x)}{(1+x^2)^2} \\
&= \frac{1 - 2x - x^2}{(1+x^2)^2}, \quad x \in \mathbb{R}_\star.
\end{aligned}
$$

Hence,

$$
\begin{aligned}
f^\star(x) &= e^{x \frac{f'(x)}{f(x)}} \\
&= e^{x \frac{\frac{1-2x-x^2}{(1+x^2)^2}}{\frac{1+x}{1+x^2}}} \\
&= e^{\frac{x(1-2x-x^2)}{(1+x)(1+x^2)}}, \quad x \in \mathbb{R}_\star.
\end{aligned}
$$

Thus,

$$f(x) > 0, \quad x \in \mathbb{R}_\star,$$

$$f^\star(x) \geq 0_\star, \quad x \in (0, -1+\sqrt{2}],$$

$$f^\star(x) \leq 0_\star, \quad x \in [-1+\sqrt{2}, \infty).$$

Thus, f is an increasing function on $(0, -1+\sqrt{2}]$, and f is a decreasing function on $[-1+\sqrt{2}, \infty)$.

Example 2.18. Let

$$f(x) = -\frac{1}{1+x}, \quad x \in \mathbb{R}_\star.$$

Then

$$f'(x) = \frac{1}{(1+x)^2}, \quad x \in \mathbb{R}_\star,$$

and

$$
\begin{aligned}
f^\star(x) &= e^{x \frac{f'(x)}{f(x)}} \\
&= e^{-x \frac{\frac{1}{(1+x)^2}}{\frac{1}{1+x}}}
\end{aligned}
$$

$$= e^{-\frac{x}{1+x}}$$

$$\leq 0_\star, \quad x \in \mathbb{R}_\star.$$

Since $f(x) < 0$, $x \in \mathbb{R}_\star$, and $f^\star(x) \leq 0$, $x \in \mathbb{R}_\star$, we conclude that f is an increasing function on $\mathbb{R}_\star$.

Exercise 2.5. Let
$$f(x) = 3x^2 + (\cos x)^2, \quad x \in \mathbb{R}_\star.$$
Prove that f is an increasing function on $\mathbb{R}_\star$.

2.6   Local Extremum

Let $A \subseteq \mathbb{R}_\star$ and $f \in \mathscr{C}^1_\star(A)$.

Theorem 2.5. *Let $x_0 \in A$, $f^\star(x_0) = 0$ and $f(x_0) \neq 0$. Then f has a local extremum at x_0.*

Proof We have

$$0_\star = 1$$

$$= e^0$$

$$= f^\star(x_0)$$

$$= e^{x_0 \frac{f'(x_0)}{f(x_0)}}.$$

Hence,
$$x_0 \frac{f'(x_0)}{f(x_0)} = 0.$$

Because $x_0 \in A$ and $x_0 \neq 0$, by the last equation, we obtain $f'(x_0) = 0$. Therefore, f has a local extremum at x_0. This completes the proof.

Theorem 2.6. *Let $x_0 \in A$ and $f^\star(x_0) = 0$, $f(x_0) \neq 0$. Let also, there is a $\delta > 0$ so that $(x_0 - \delta, x_0 + \delta) \subset A$ and*

$$f(x) > 0, \quad x \in (x_0 - \delta, x_0 + \delta),$$

$$f^\star(x) > 0_\star, \quad x \in (x_0 - \delta, x_0),$$

$$f^\star(x) \quad < \quad 0_\star, \quad x \in (x_0, x_0 + \delta).$$

Then f has a local maximum at x_0.

Proof Since $f^\star(x_0) = 0$, we have that f has a local extremum at x_0. Because

$$f^\star(x) = e^{x\frac{f'(x)}{f(x)}}, \quad x \in (x_0 - \delta, x_0 + \delta),$$

and $xf(x) > 0$, $x \in (x_0 - \delta, x_0 + \delta)$, we have that

$$f^\star(x) > 0_\star, \quad x \in (x_0 - \delta, x_0), \quad \text{iff} \quad f'(x) > 0, \quad x \in (x_0 - \delta, x_0),$$

and

$$f^\star(x) < 0_\star, \quad x \in (x_0, x_0 + \delta), \quad \text{iff} \quad f'(x) < 0, \quad x \in (x_0, x_0 + \delta).$$

So, we conclude that f has a local maximum at x_0. This completes the proof.

Theorem 2.7. *Let $x_0 \in A$ and $f^\star(x_0) = 0$, $f(x_0) \neq 0$. Let also, there is a $\delta > 0$ so that $(x_0 - \delta, x_0 + \delta) \subset A$ and*

$$f(x) \quad < \quad 0, \quad x \in (x_0 - \delta, x_0 + \delta),$$

$$f^\star(x) \quad > \quad 0_\star, \quad x \in (x_0 - \delta, x_0),$$

$$f^\star(x) \quad < \quad 0_\star, \quad x \in (x_0, x_0 + \delta).$$

Then f has a local minimum at x_0.

Proof Since $f^\star(x_0) = 0$, we have that f has a local extremum at x_0. Because

$$f^\star(x) = e^{x\frac{f'(x)}{f(x)}}, \quad x \in (x_0 - \delta, x_0 + \delta),$$

and $xf(x) < 0$, $x \in (x_0 - \delta, x_0 + \delta)$, we have that

$$f^\star(x) > 0_\star, \quad x \in (x_0 - \delta, x_0), \quad \text{iff} \quad f'(x) < 0, \quad x \in (x_0 - \delta, x_0),$$

and

$$f^\star(x) < 0_\star, \quad x \in (x_0, x_0 + \delta), \quad \text{iff} \quad f'(x) > 0, \quad x \in (x_0, x_0 + \delta).$$

So, we conclude that f has a local minimum at x_0. This completes the proof.

Theorem 2.8. *Let $x_0 \in A$ and $f^\star(x_0) = 0$, $f(x_0) \neq 0$. Let also, there is a $\delta > 0$ so that $(x_0 - \delta, x_0 + \delta) \subset A$ and*

$$f(x) \quad > \quad 0, \quad x \in (x_0 - \delta, x_0 + \delta),$$

$$f^\star(x) \quad < \quad 0_\star, \quad x \in (x_0 - \delta, x_0),$$

$$f^\star(x) \quad > \quad 0_\star, \quad x \in (x_0, x_0 + \delta).$$

Then f has a local minimum at x_0.

Proof Since $f^\star(x_0) = 0$, we have that f has a local extremum at x_0. Because

$$f^\star(x) = e^{x\frac{f'(x)}{f(x)}}, \quad x \in (x_0 - \delta, x_0 + \delta),$$

and $xf(x) > 0$, $x \in (x_0 - \delta, x_0 + \delta)$, we have that

$$f^\star(x) < 0_\star, \quad x \in (x_0 - \delta, x_0), \quad \text{iff} \quad f'(x) < 0, \quad x \in (x_0 - \delta, x_0),$$

and

$$f^\star(x) > 0_\star, \quad x \in (x_0, x_0 + \delta), \quad \text{iff} \quad f'(x) > 0, \quad x \in (x_0, x_0 + \delta).$$

So, we conclude that f has a local minimum at x_0. This completes the proof.

Theorem 2.9. *Let $x_0 \in A$ and $f^\star(x_0) = 0$, $f(x_0) \neq 0$. Let also, there is a $\delta > 0$ so that $(x_0 - \delta, x_0 + \delta) \subset A$ and*

$$f(x) \quad < \quad 0, \quad x \in (x_0 - \delta, x_0 + \delta),$$

$$f^\star(x) \quad < \quad 0_\star, \quad x \in (x_0 - \delta, x_0),$$

$$f^\star(x) \quad > \quad 0_\star, \quad x \in (x_0, x_0 + \delta).$$

Then f has a local maximum at x_0.

Proof Since $f^\star(x_0) = 0$, we have that f has a local extremum at x_0. Because

$$f^\star(x) = e^{x\frac{f'(x)}{f(x)}}, \quad x \in (x_0 - \delta, x_0 + \delta),$$

and $xf(x) < 0$, $x \in (x_0 - \delta, x_0 + \delta)$, we have that

$$f^\star(x) < 0_\star, \quad x \in (x_0 - \delta, x_0), \quad \text{iff} \quad f'(x) > 0, \quad x \in (x_0 - \delta, x_0),$$

and

$$f^\star(x) > 0_\star, \quad x \in (x_0, x_0 + \delta), \quad \text{iff} \quad f'(x) < 0, \quad x \in (x_0, x_0 + \delta).$$

So, we conclude that f has a local maximum at x_0. This completes the proof.

Theorem 2.10. *Let $x_0 \in A$ and $f^\star(x_0) = 0$, $f(x_0) \neq 0$. Let also, there is a $\delta > 0$ so that $(x_0 - \delta, x_0 + \delta) \subset A$ and*

$$f^\star(x) \quad > \quad 0_\star, \quad x \in (x_0 - \delta, x_0 + \delta),$$

$$f(x) \quad > \quad 0, \quad x \in (x_0 - \delta, x_0),$$

$$f(x) \quad < \quad 0, \quad x \in (x_0, x_0 + \delta).$$

Then f has a local maximum at x_0.

Proof Since $f^\star(x_0) = 0$, we have that f has a local extremum at x_0. Because

$$f^\star(x) = e^{x \frac{f'(x)}{f(x)}}, \quad x \in (x_0 - \delta, x_0 + \delta),$$

and $f^\star(x) > 0_\star, x \in (x_0 - \delta, x_0 + \delta)$, we have that

$$f(x) > 0, \quad x \in (x_0 - \delta, x_0), \quad \text{iff} \quad f'(x) > 0, \quad x \in (x_0 - \delta, x_0),$$

and

$$f(x) < 0, \quad x \in (x_0, x_0 + \delta), \quad \text{iff} \quad f'(x) < 0, \quad x \in (x_0, x_0 + \delta).$$

So, we conclude that f has a local maximum at x_0. This completes the proof.

Theorem 2.11. *Let $x_0 \in A$ and $f^\star(x_0) = 0$, $f(x_0) \neq 0$. Let also, there is a $\delta > 0$ so that $(x_0 - \delta, x_0 + \delta) \subset A$ and*

$$f^\star(x) \quad > \quad 0_\star, \quad x \in (x_0 - \delta, x_0 + \delta),$$

$$f(x) \quad < \quad 0, \quad x \in (x_0 - \delta, x_0),$$

$$f(x) \quad > \quad 0, \quad x \in (x_0, x_0 + \delta).$$

Then f has a local minimum at x_0.

Proof Since $f^\star(x_0) = 0$, we have that f has a local extremum at x_0. Because

$$f^\star(x) = e^{x \frac{f'(x)}{f(x)}}, \quad x \in (x_0 - \delta, x_0 + \delta),$$

and $f^\star(x) > 0_\star, x \in (x_0 - \delta, x_0 + \delta)$, we have that

$$f(x) < 0, \quad x \in (x_0 - \delta, x_0), \quad \text{iff} \quad f'(x) < 0, \quad x \in (x_0 - \delta, x_0),$$

and

$$f(x) > 0, \quad x \in (x_0, x_0 + \delta), \quad \text{iff} \quad f'(x) > 0, \quad x \in (x_0, x_0 + \delta).$$

So, we conclude that f has a local minimum at x_0. This completes the proof.

Theorem 2.12. *Let $x_0 \in A$ and $f^\star(x_0) = 0$, $f(x_0) \neq 0$. Let also, there is a $\delta > 0$ so that $(x_0 - \delta, x_0 + \delta) \subset A$ and*

$$f^\star(x) \quad < \quad 0_\star, \quad x \in (x_0 - \delta, x_0 + \delta),$$

$$f(x) \quad > \quad 0, \quad x \in (x_0 - \delta, x_0),$$

$$f(x) \quad < \quad 0, \quad x \in (x_0, x_0 + \delta).$$

Then f has a local minimum at x_0.

Proof Since $f^\star(x_0) = 0$, we have that f has a local extremum at x_0. Because

$$f^\star(x) = e^{x\frac{f'(x)}{f(x)}}, \quad x \in (x_0 - \delta, x_0 + \delta),$$

and $f^\star(x) < 0_\star$, $x \in (x_0 - \delta, x_0 + \delta)$, we have that

$$f(x) > 0, \quad x \in (x_0 - \delta, x_0), \quad \text{iff} \quad f'(x) < 0, \quad x \in (x_0 - \delta, x_0),$$

and

$$f(x) < 0, \quad x \in (x_0, x_0 + \delta), \quad \text{iff} \quad f'(x) > 0, \quad x \in (x_0, x_0 + \delta).$$

So, we conclude that f has a local minimum at x_0. This completes the proof.

Theorem 2.13. *Let* $x_0 \in A$ *and* $f^\star(x_0) = 0$, $f(x_0) \neq 0$. *Let also, there is a* $\delta > 0$ *so that* $(x_0 - \delta, x_0 + \delta) \subset A$ *and*

$$f^\star(x) \quad < \quad 0_\star, \quad x \in (x_0 - \delta, x_0 + \delta),$$

$$f(x) \quad < \quad 0, \quad x \in (x_0 - \delta, x_0),$$

$$f(x) \quad > \quad 0, \quad x \in (x_0, x_0 + \delta).$$

Then f *has a local maximum at* x_0.

Proof Since $f^\star(x_0) = 0$, we have that f has a local extremum at x_0. Because

$$f^\star(x) = e^{x\frac{f'(x)}{f(x)}}, \quad x \in (x_0 - \delta, x_0 + \delta),$$

and $f^\star(x) < 0_\star$, $x \in (x_0 - \delta, x_0 + \delta)$, we have that

$$f(x) < 0, \quad x \in (x_0 - \delta, x_0), \quad \text{iff} \quad f'(x) > 0, \quad x \in (x_0 - \delta, x_0),$$

and

$$f(x) > 0, \quad x \in (x_0, x_0 + \delta), \quad \text{iff} \quad f'(x) < 0, \quad x \in (x_0, x_0 + \delta).$$

So, we conclude that f has a local maximum at x_0. This completes the proof.

2.7   The Multiplicative Rolle Theorem

Let $a, b \in \mathbb{R}_\star$, $a < b$.

Theorem 2.14 (The Multiplicative Rolle Theorem). *Let* $f \in \mathscr{C}_\star^1((a,b)) \cap \mathscr{C}_\star([a,b])$ *and* $f(a) = f(b)$. *Then there exists a* $c \in (a,b)$ *so that* $f^\star(c) = 0_\star$.

Proof By the classical Rolle theorem, it follows that there is a $c \in (a,b)$ so that $f'(c) = 0$. Hence,

$$f^\star(c) = e^{c\frac{f'(c)}{f(c)}}$$

$$= e^0$$

$$= 1$$

$$= 0_\star.$$

This completes the proof. ∎

2.8 The Multiplicative Lagrange Theorem

Let $a, b \in \mathbb{R}_\star$, $a < b$.

Theorem 2.15 (The Multiplicative Lagrange Theorem). *Let $f \in \mathscr{C}_\star^1((a,b)) \cap \mathscr{C}_\star([a,b])$. Then there is a $c \in (a,b)$ so that*

$$(f(b) -_\star f(a))/_\star(b -_\star a) = f^\star(c).$$

Proof For $t_1, t_2 \in [a,b]$, we have

$$f(t_1) -_\star f(t_2) = \frac{f(t_1)}{f(t_2)},$$

$$t_1 -_\star t_2 = \frac{t_1}{t_2}.$$

Then,

$$(f(t_1) -_\star f(t_2))/_\star(t_1 -_\star t_2) = \left(\frac{f(t_1)}{f(t_2)}\right)/_\star\left(\frac{t_1}{t_2}\right)$$

$$= e^{\frac{\log \frac{f(t_1)}{f(t_2)}}{\log \frac{t_1}{t_2}}}$$

$$= e^{\frac{\log f(t_1) - \log f(t_2)}{\log t_1 - \log t_2}}.$$

In particular, we have

$$(f(b) -_\star f(a))/_\star(b -_\star a) = e^{\frac{\log f(b) - \log f(a)}{\log b - \log a}}.$$

By the classical Cauchy theorem, it follows that there is a $c \in (a,b)$ so that

$$\frac{\log f(b) - \log f(a)}{\log b - \log a} = \frac{\frac{f'(c)}{f(c)}}{\frac{1}{c}}$$

$$= c\frac{f'(c)}{f(c)}.$$

Hence,

$$(f(b) -_\star f(a))/_\star(b -_\star a) = e^{\frac{\log f(b) - \log f(a)}{\log b - \log a}}$$

$$= e^{c\frac{f'(c)}{f(c)}}$$

$$= f^\star(c).$$

This completes the proof.

2.9   The Multiplicative Cauchy Theorem

Let $a, b \in \mathbb{R}_\star$, $a < b$.

Theorem 2.16 (The Multiplicative Cauchy Theorem). *Let* $f, g \in \mathscr{C}_\star^1((a,b)) \cap \mathscr{C}_\star([a,b])$. *Then there is a* $c \in (a,b)$ *so that*

$$(f(b) -_\star f(a))/_\star(g(b) -_\star g(a)) = (f^\star(c))/_\star(g^\star(c)).$$

Proof For $t_1, t_2 \in [a,b]$, we have

$$f(t_1) -_\star f(t_2) = \frac{f(t_1)}{f(t_2)},$$

$$g(t_1) -_\star g(t_2) = \frac{g(t_1)}{g(t_2)}.$$

Then

$$(f(t_1) -_\star f(t_2))/_\star(g(t_1) -_\star g(t_2)) = \left(\frac{f(t_1)}{f(t_2)}\right)/_\star\left(\frac{g(t_1)}{g(t_2)}\right)$$

$$= e^{\frac{\log\frac{f(t_1)}{f(t_2)}}{\log\frac{g(t_1)}{g(t_2)}}}$$

$$= e^{\frac{\log f(t_1) - \log f(t_2)}{\log g(t_1) - \log g(t_2)}}.$$

In particular, we have

$$(f(b) -_\star f(a)) /_\star (g(b) -_\star g(a)) = e^{\frac{\log f(b) - \log f(a)}{\log g(b) - \log g(a)}}.$$

By the classical Cauchy theorem, it follows that there is a $c \in (a, b)$ so that

$$\frac{\log f(b) - \log f(a)}{\log g(b) - \log g(a)} = \frac{\frac{f'(c)}{f(c)}}{\frac{g'(c)}{g(c)}}$$

$$= \frac{c\frac{f'(c)}{f(c)}}{c\frac{g'(c)}{g(c)}}.$$

Hence,

$$(f(b) -_\star f(a)) /_\star (g(b) -_\star g(a)) = e^{\frac{\log f(b) - \log f(a)}{\log g(b) - \log g(a)}}$$

$$= e^{\frac{c\frac{f'(c)}{f(c)}}{c\frac{g'(c)}{g(c)}}}$$

$$= e^{c\frac{f'(c)}{f(c)}} /_\star e^{c\frac{g'(c)}{g(c)}}$$

$$= f^\star(c) /_\star g^\star(c).$$

This completes the proof.

2.10 The Multiplicative Taylor Formula

Let $a, b \in \mathbb{R}_\star$, $a < b$, and $f : \mathbb{R}_\star \to \mathbb{R}_\star$ is enough times multiplicative differentiable on $[a, b]$. Take

$$a_j = f_\star^{\star(j)}(x_0) /_\star j!_\star, \quad j \in \{0, 1, \ldots, n\}.$$

Define the polynomial

$$P_n(x, x_0) = a_n \cdot_\star (x -_\star x_0)^{n_\star} +_\star a_{n-1} \cdot_\star (x -_\star x_0)^{n-1_\star}$$

$$+_\star \cdots +_\star a_1 \cdot_\star (x -_\star x_0) +_\star a_0, \quad x \in [a, b].$$

We have

$$P_n(x, x_0) = a_n \cdot_\star \left(\frac{x}{x_0}\right)^{n_\star} +_\star a_{n-1} \cdot_\star \left(\frac{x}{x_0}\right)^{n-1_\star} +_\star \cdots +_\star a_1 \cdot_\star \frac{x}{x_0} +_\star a_0$$

$$= a_n \cdot_\star e^{\left(\log\frac{x}{x_0}\right)^n} +_\star a_{n-1} \cdot_\star e^{\left(\log\frac{x}{x_0}\right)^{n-1}} +_\star \cdots +_\star a_1 \cdot_\star e^{\log\frac{x}{x_0}} +_\star a_0$$

$$= e^{\log a_n \left(\log\frac{x}{x_0}\right)^n + \log a_{n-1}\left(\log\frac{x}{x_0}\right)^{n-1} + \cdots + \log a_1 \log\frac{x}{x_0} + \log a_0}.$$

We apply the classical Taylor formula for the function $f(\log x)$ and we get that there is a $c \in (a,b)$ so that

$$f_\star(x) = e^{f(\log x)}$$

$$= e^{\sum_{j=1}^{n} \log a_j \left(\log\frac{x}{x_0}\right)^j + \log\left(f_\star^{\star(n+1)}(c)/_\star(n+1)!_\star\right)\left(\log\frac{x}{x_0}\right)^{n+1}}, \tag{2.1}$$

$x \in [a,b]$.

Definition 2.5. The formula (2.1) will be called the multiplicative Taylor formula.

2.11   Advanced Practical Problems

Problem 2.1. Let

1. $f(x) = 2 + x^4$, $x \in \mathbb{R}_\star$.

2. $f(x) = \dfrac{1+x^2}{2+x^2}$, $x \in \mathbb{R}_\star$.

3. $f(x) = \tan x + x^4 + 20$, $x \in \left(\dfrac{1}{3}, \dfrac{1}{2}\right)$.

Find $f^\star(x)$, $x \in \mathbb{R}_\star$.

Answer 2.5.

1. $f_\star^\star(x) = e^{\frac{4x^4}{2+x^4}}$, $x \in \mathbb{R}_\star$.

2. $f_\star^\star(x) = e^{\frac{2x^2}{(1+x^2)(2+x^2)}}$, $x \in \mathbb{R}_\star$.

3. $f_\star^\star(x) = e^{\frac{x\left(4x^3 + \frac{1}{(\cos x)^2}\right)}{\tan x + x^4 + 20}}$, $x \in \left(\dfrac{1}{3}, \dfrac{1}{2}\right)$.

Problem 2.2. Let

1. $f(x) = \cos(2x)$, $x \in \mathbb{R}_\star$.
2. $f(x) = 2 - 3x - x^2$, $x \in \mathbb{R}_\star$.
3. $f(x) = \sin(4x)$, $x \in \mathbb{R}_\star$.

Find $f_\star^\star(x)$, $x \in \mathbb{R}_\star$.

Answer 2.6.

1. $f_\star^\star(x) = e^{-2\sin(2\log x)}$, $x \in \mathbb{R}_\star$.
2. $f_\star^\star(x) = e^{-3-2\log x}$, $x \in \mathbb{R}_\star$.
3. $f_\star^\star(x) = e^{4\cos(4\log x)}$, $x \in \mathbb{R}_\star$.

Problem 2.3. Let $f(x) = x^3$, $x \in \mathbb{R}_\star$. Find

$$f^\star(x), \quad f^{\star\star}(x), \quad x \in \mathbb{R}_\star.$$

Answer 2.7.

1. $f^\star(x) = e^3$, $x \in \mathbb{R}_\star$.
2. $f^{\star\star}(x) = 1$, $x \in \mathbb{R}_\star$.

Problem 2.4. Let

$$f(x) = x^3 - \frac{1}{1+x}, \quad x \in \mathbb{R}_\star.$$

Prove that f is an increasing function on $\mathbb{R}_\star$.

Problem 2.5. Let

$$x(t) \quad = \quad \frac{1}{3+t},$$

$$y(t) \quad = \quad \frac{1}{4+t}, \quad t \in \mathbb{R}_\star.$$

Find

$$d_\star^2 y / {}_\star d_\star x^{2\star}.$$

Answer 2.8.

$$e^{-\frac{3+t}{(4+t)^2}}, \quad t \in \mathbb{R}_\star.$$

3

Multiplicative Integration

In this chapter, we introduce the indefinite multiplicative integral and the Cauchy multiplicative integral and deduct some of their properties. We deduct the table of the basic multiplicative integrals. We consider multiplicative integration by substitutions and multiplicative integration by parts. We prove some important inequalities for the multiplicative integrals and we prove some mean value theorems for multiplicative integrals.

3.1 Definition for the Multiplicative Improper Integral and Multiplicative Cauchy Integral

Suppose that $a, b \in \mathbb{R}_\star$ and $f : \mathbb{R}_\star \to \mathbb{R}_\star$, $f \in \mathscr{C}(\mathbb{R}_\star)$.

Definition 3.1. The integral,

$$\int_\star f(x) \cdot_\star d_\star x = e^{\int \frac{1}{x} \log f(x) dx}, \quad x \in \mathbb{R}_\star,$$

will be called multiplicative indefinite integral. Define the multiplicative Cauchy integral as follows

$$\int_{\star a}^b f(s) \cdot_\star d_\star x = e^{\int_a^b \frac{1}{s} \log f(s) ds}.$$

For $x \in \mathbb{R}_\star$, we have

$$
\begin{aligned}
d_\star \left(\int_{\star a}^x f(s) \cdot_\star d_\star s \right) /_\star d_\star x &= d_\star \left(e^{\int_a^x \frac{1}{s} \log f(s) ds} \right) /_\star d_\star x \\[2mm]
&= e^{x \frac{\left(e^{\int_a^x \frac{1}{s} \log f(s) ds} \right)'}{e^{\int_a^x \frac{1}{s} \log f(s) ds}}} \\[2mm]
&= e^{x \frac{1}{x} \log f(x)} \\[2mm]
&= e^{\log f(x)} \\[2mm]
&= f(x), \quad x \in \mathbb{R}_\star.
\end{aligned}
$$

DOI: 10.1201/9781003299080-3

Moreover,

$$\int_{\star a}^{b} f^{\star}(s) \cdot_{\star} d_{\star}s \quad = \quad e^{\int_a^b \frac{1}{s} \log f^{\star}(s) ds}$$

$$= \quad e^{\int_a^b \frac{1}{s} \log e^{\frac{s f'(s)}{f(s)}} ds}$$

$$= \quad e^{\int_a^b \frac{1}{s} s \frac{f'(s)}{f(s)} ds}$$

$$= \quad e^{\int_a^b \frac{f'(s)}{f(s)} ds}$$

$$= \quad e^{\int_a^b d \log f(s)}$$

$$= \quad e^{\log f(b) - \log f(a)}$$

$$= \quad e^{\log \frac{f(b)}{f(a)}}$$

$$= \quad \frac{f(b)}{f(a)}$$

$$= \quad f(b) -_{\star} f(a).$$

Example 3.1. Let $f(x) = e^{x \sin x}$, $x \in \mathbb{R}_{\star}$. Then

$$\int_{\star \frac{\pi}{4}}^{\frac{\pi}{2}} f(x) \cdot_{\star} d_{\star}x \quad = \quad e^{\int_{\frac{\pi}{4}}^{\frac{\pi}{2}} \frac{1}{x} \log e^{x \sin x} dx}$$

$$= \quad e^{\int_{\frac{\pi}{4}}^{\frac{\pi}{2}} \sin x dx}$$

$$= \quad e^{-\cos x \Big|_{x=\frac{\pi}{4}}^{x=\frac{\pi}{2}}}$$

$$= \quad e^{\cos \frac{\pi}{4}}$$

$$= \quad e^{\frac{\sqrt{2}}{2}}.$$

Example 3.2. Let $f(x) = e^{x^2 + x^3}$, $x \in \mathbb{R}_{\star}$. Then

$$\int_{\star 2}^{3} f(x) \cdot_{\star} d_{\star}x \quad = \quad e^{\int_2^3 \frac{1}{x} \log f(x) dx}$$

$$= e^{\int_2^3 \frac{1}{x} \log e^{x^2 + x^3} dx}$$

$$= e^{\int_2^3 \frac{1}{x}(x^2 + x^3)dx}$$

$$= e^{\int_2^3 (x + x^2)dx}$$

$$= e^{\left. \frac{1}{2}x^2 \right|_{x=2}^{x=3} + \left. \frac{1}{3}x^3 \right|_{x=2}^{x=3}}$$

$$= e^{\frac{1}{2}(9-4) + \frac{1}{3}(27-8)}$$

$$= e^{\frac{5}{2} + \frac{19}{3}}$$

$$= e^{\frac{15+38}{6}}$$

$$= e^{\frac{53}{6}}.$$

Example 3.3. Let $f(x) = e^{\frac{1}{x^4}}$, $x \in \mathbb{R}_\star$. Then

$$\int_{\star 1}^{2} f(x) \cdot_\star d_\star x = e^{\int_1^2 \frac{1}{x} \log e^{\frac{1}{x^4}} dx}$$

$$= e^{\int_1^2 \frac{1}{x^5} dx}$$

$$= e^{\left. -\frac{1}{4x^4} \right|_{x=1}^{x=2}}$$

$$= e^{-\frac{1}{64} + \frac{1}{4}}$$

$$= e^{\frac{15}{64}}.$$

Exercise 3.1. Let $f(x) = e^{x^3 + x}$, $x \in \mathbb{R}_\star$. Find

$$\int_{\star 1}^{2} f(x) \cdot_\star d_\star x.$$

Answer 3.1. $e^{\frac{10}{3}}$.

3.2   Table of the Basic Multiplicative Integrals

In this section, we will deduct the basic multiplicative integrals.

1. Let $f(x) = x^{k_\star}$, $k \neq -1$. Then

$$f(x) = e^{(\log x)^k}, \quad x \in \mathbb{R}_\star,$$

and

$$
\begin{aligned}
\int_\star f(x) \cdot_\star d_\star x &= e^{\int \frac{1}{x} \log f(x) dx} \\[2mm]
&= e^{\int \frac{1}{x} (\log x)^k dx} \\[2mm]
&= e^{\int (\log x)^k d \log x} \\[2mm]
&= c e^{\frac{1}{k+1} (\log x)^{k+1}}, \quad x \in \mathbb{R}_\star,
\end{aligned}
$$

where c is a positive constant.

2. Let $a \in \mathbb{R}_\star$. Then

$$
\begin{aligned}
\int_\star a \cdot_\star d_\star x &= e^{\int \frac{1}{x} \log a \, dx} \\[2mm]
&= c e^{\log a \log x} \\[2mm]
&= c x^{\log a}, \quad x \in \mathbb{R}_\star,
\end{aligned}
$$

where c is a positive constant.

3. Let $a > 0$. We have

$$
\begin{aligned}
\int_\star (a^x)_\star \cdot_\star d_\star x &= e^{\int \frac{1}{x} a^{\log x} dx} \\[2mm]
&= e^{\int a^{\log x} d \log x} \\[2mm]
&= c e^{\frac{a^{\log x}}{\log a}}, \quad x \in \mathbb{R}_\star, \quad a > 0.
\end{aligned}
$$

4. We have

$$
\int_\star \sin_\star x \cdot_\star d_\star x = e^{\int \frac{1}{x} \sin(\log x) dx}
$$

$$= \quad e^{\int \sin(\log x) d\log x}$$

$$= \quad ce^{-\cos(\log x)}, \quad x \in \mathbb{R}_\star, \quad c > 0.$$

5. We have

$$\int_\star \cos_\star x \cdot_\star d_\star x \quad = \quad e^{\int \frac{1}{x} \cos(\log x) dx}$$

$$= \quad e^{\int \cos(\log x) d\log x}$$

$$= \quad ce^{\sin(\log x)}, \quad x \in \mathbb{R}_\star, \quad c > 0.$$

6. We have

$$\int_\star (1_\star /_\star (\cos_\star x)^{2_\star}) \cdot_\star d_\star x \quad = \quad \int_\star e^{\frac{1}{(\cos(\log x))^2}} \cdot_\star d_\star x$$

$$= \quad e^{\int \frac{1}{x} \frac{1}{(\cos(\log x))^2} dx}$$

$$= \quad e^{\int \frac{1}{(\cos(\log x))^2} d\log x}$$

$$= \quad ce^{\tan(\log x)}, \quad x \in \mathbb{R}_\star, \quad c > 0.$$

7. We have

$$\int_\star (1_\star /_\star (\sin_\star x)^{2_\star}) \cdot_\star d_\star x \quad = \quad \int_\star e^{\frac{1}{(\sin(\log x))^2}} \cdot_\star d_\star x$$

$$= \quad e^{\int \frac{1}{x} \frac{1}{(\sin(\log x))^2} dx}$$

$$= \quad e^{\int \frac{1}{(\sin(\log x))^2} d\log x}$$

$$= \quad ce^{-\cot(\log x)}, \quad x \in \mathbb{R}_\star, \quad c > 0.$$

8. We have

$$\int_\star (1_\star /_\star x_\star) \cdot_\star d_\star x \quad = \quad \int_\star e^{\frac{1}{\log x}} \cdot_\star d_\star x$$

$$= \quad e^{\int \frac{1}{x} \log e^{\frac{1}{\log x}} dx}$$

$$= e^{\int \frac{1}{x \log x} dx}$$

$$= e^{\int \frac{1}{\log x} d \log x}$$

$$= c_1 e^{\log |\log x|}$$

$$= c \log x, \quad c_1 > 0, \quad c \in \mathbb{R}, \quad x \in \mathbb{R}_\star.$$

9. We have

$$\int_\star \left(1_\star /_\star (1_\star +_\star x_\star^{2\star}) \right) \cdot_\star d_\star x = \int_\star \left(1_\star /_\star (e x_\star^{2\star}) \right) \cdot_\star d_\star x$$

$$= e^{\int \frac{1}{x} \log e^{\frac{1}{1+(\log x)^2}} dx}$$

$$= e^{\int \frac{1}{1+(\log x)^2} d \log x}$$

$$= c e^{\arctan(\log x)}, \quad x \in \mathbb{R}_\star, \quad c > 0.$$

10. We have

$$\int_\star \left(1_\star /_\star (1_\star -_\star x_\star^{2\star})^{\frac{1}{2}\star} \right) \cdot_\star d_\star x = \int_\star \left(e /_\star \left(\frac{e}{x_\star^{2\star}} \right) \right)^{\frac{1}{2}\star} \cdot_\star d_\star x$$

$$= \int_\star \left(e /_\star \left(\frac{e}{e^{(\log x)^2}} \right) \right)^{\frac{1}{2}\star} \cdot_\star d_\star x$$

$$= \int_\star \left(e /_\star \left(e^{1-(\log x)^2} \right) \right)^{\frac{1}{2}\star} \cdot_\star d_\star x$$

$$= \int_\star \left(e^{\frac{1}{1-(\log x)^2}} \right)^{\frac{1}{2}\star} \cdot_\star d_\star x$$

$$= \int_\star e^{\sqrt{\frac{1}{1-(\log x)^2}}} \cdot_\star d_\star x$$

$$= e^{\int \frac{1}{x} \log e^{\sqrt{\frac{1}{1-(\log x)^2}}} dx}$$

$$= e^{\int \frac{1}{\sqrt{1-(\log x)^2}} d \log x}$$

$$= c e^{-\arccos(\log x)}$$

$$= c e^{\arcsin(\log x)}, \quad x \in \mathbb{R}_\star, \quad c > 0.$$

11. Let $a > 0$. Then

$$\int_\star \left(1_\star /_\star \left(x_\star^{2\star} +_\star a^{2\star} \right)^{\frac{1}{2}\star} \right) \cdot_\star d_\star x \;=\; \int_\star \left(e /_\star \left(x_\star^{2\star} a^{2\star} \right)^{\frac{1}{2}\star} \right) \cdot_\star d_\star x$$

$$= \int_\star \left(e /_\star \left(e^{(\log x)^2 + (\log a)^2} \right)^{\frac{1}{2}\star} \right) \cdot_\star d_\star x$$

$$= \int_\star \left(e /_\star \left(e^{\left((\log x)^2 + (\log a)^2 \right)^{\frac{1}{2}}} \right) \right) \cdot_\star d_\star x$$

$$= \int_\star e^{\sqrt{\dfrac{1}{(\log x)^2 + (\log a)^2}}} \cdot_\star d_\star x$$

$$= e^{\int \frac{1}{x} \log e \sqrt{\frac{1}{(\log x)^2 + (\log a)^2}} \, dx}$$

$$= e^{\int \frac{1}{\sqrt{(\log x)^2 + (\log a)^2}} \, d\log x}$$

$$= c \, e^{\log \left(\log x + \sqrt{(\log x)^2 + (\log a)^2} \right)}$$

$$= c \left(\log x + \sqrt{(\log x)^2 + (\log a)^2} \right),$$

$$x \in \mathbb{R}_\star, \quad c > 0.$$

12. Let $a > 0$. Then

$$\int_\star \left(1_\star /_\star \left(x_\star^{2\star} -_\star a^{2\star} \right)^{\frac{1}{2}\star} \right) \cdot_\star d_\star x \;=\; \int_\star \left(e /_\star \left(\frac{x_\star^{2\star}}{a^{2\star}} \right)^{\frac{1}{2}\star} \right) \cdot_\star d_\star x$$

$$= \int_\star \left(e /_\star \left(\frac{e^{(\log x)^2}}{e^{(\log a)^2}} \right)^{\frac{1}{2}\star} \right) \cdot_\star d_\star x$$

$$= \int_\star \left(e /_\star \left(e^{(\log x)^2 - (\log a)^2} \right)^{\frac{1}{2}\star} \right) \cdot_\star d_\star x$$

$$= \int_\star \left(e /_\star e^{\sqrt{(\log x)^2 - (\log a)^2}} \right) \cdot_\star d_\star x$$

$$= e^{\int \frac{1}{x} e^{\sqrt{\frac{1}{(\log x)^2 - (\log a)^2}}} \, dx}$$

$$= e^{\int \frac{1}{\sqrt{(\log x)^2 - (\log a)^2}} d\log x}$$

$$= c_1 e^{\log\left|\log x + \sqrt{(\log x)^2 - (\log a)^2}\right|}$$

$$= c\left(\log x + \sqrt{(\log x)^2 - (\log a)^2}\right),$$

$$x \in \mathbb{R}_\star, \quad c_1 > 0, \quad c \in \mathbb{R}.$$

13. Let $a > 0$. Then

$$\int_\star \left(1_\star /_\star \left(x_\star^{2\star} -_\star a^{2\star}\right)\right) \cdot_\star d_\star x = \int_\star \left(1_\star /_\star \left(\frac{x_\star^{2\star}}{a^{2\star}}\right)\right) \cdot_\star d_\star x$$

$$= \int_\star \left(e/_\star \frac{e^{(\log x)^2}}{e^{(\log a)^2}}\right) \cdot_\star d_\star x$$

$$= \int_\star \left(e/_\star e^{(\log x)^2 - (\log a)^2}\right) \cdot_\star d_\star x$$

$$= \int_\star e^{\frac{1}{(\log x)^2 - (\log a)^2}} \cdot_\star d_\star x$$

$$= e^{\int \frac{1}{x} \log e^{\frac{1}{(\log x)^2 - (\log a)^2}} dx}$$

$$= e^{\int \frac{1}{(\log x)^2 - (\log a)^2} d\log x}$$

$$= c e^{\frac{1}{2\log a} \left|\frac{\log x - \log a}{\log x + \log a}\right|}$$

$$= c \left|\frac{\log x - \log a}{\log x + \log a}\right|^{\frac{1}{2\log a}}$$

$$= c \left|\frac{\log\left(\frac{x}{a}\right)}{\log(xa)}\right|^{\frac{1}{2\log a}}, \quad x \in \mathbb{R}_\star, \quad c > 0.$$

14. We have

$$\int_\star \sinh_\star x \cdot_\star d_\star x = \int_\star e^{\sinh(\log x)} \cdot_\star d_\star x$$

$$= e^{\int \frac{1}{x} \sinh(\log x) dx}$$

$$= e^{\int \sinh(\log x)\, d\log x}$$

$$= c e^{\cosh(\log x)}, \quad x \in \mathbb{R}_\star, \quad c > 0.$$

15. We have

$$\int_\star \cosh_\star x \cdot_\star d_\star x = \int_\star e^{\cosh(\log x)} \cdot_\star d_\star x$$

$$= e^{\int \frac{1}{x} \cosh(\log x)\, dx}$$

$$= e^{\int \cosh(\log x)\, d\log x}$$

$$= c e^{\sinh(\log x)}, \quad x \in \mathbb{R}_\star, \quad c > 0.$$

3.3 Properties of the Multiplicative Integrals

Suppose that $a, b, c \in \mathbb{R}_\star$, $f, g : \mathbb{R}_\star \to \mathbb{R}_\star$, $f, g \in \mathscr{C}_\star(\mathbb{R}_\star)$.

1. We have

$$\int_\star (a \cdot_\star f(x) +_\star b \cdot_\star g(x)) \cdot_\star d_\star x = a \cdot_\star \int_\star f(x) \cdot_\star d_\star x +_\star b \cdot_\star \int_\star g(x) \cdot_\star d_\star x.$$

Proof Note that

$$a \cdot_\star f(x) +_\star b \cdot_\star g(x) = e^{\log a \log f(x)} +_\star e^{\log b \log g(x)}$$

$$= e^{\log a \log f(x) + \log b \log g(x)}, \quad x \in \mathbb{R}_\star.$$

Hence,

$$\int_\star (a \cdot_\star f(x) +_\star b \cdot_\star g(x)) \cdot_\star d_\star x = \int_\star e^{\log a \log f(x) + \log b \log gx} \cdot_\star d_\star x$$

$$= e^{\int \frac{1}{x} \log e^{\log a \log f(x) + \log b \log g(x)}\, dx}$$

$$= e^{\int \frac{1}{x} \log a \log f(x)\, dx + \int \frac{1}{x} \log b \log g(x)\, dx}$$

$$= e^{\log a \int \frac{1}{x} \log f x\, dx + \log b \int \frac{1}{x} \log g(x)\, dx}$$

$$= \quad e^{\log a \log e^{\int \frac{1}{x} \log x dx}} e^{\log b \log e^{\int \frac{1}{x} \log g(x) dx}}$$

$$= \quad \left(a \cdot_\star e^{\int \frac{1}{x} \log f(x) dx} \right) \left(b \cdot_\star e^{\int \frac{1}{x} \log g(x) dx} \right)$$

$$= \quad \left(a \cdot_\star \int_\star f(x) \cdot_\star d_\star x \right) \left(b \cdot_\star \int_\star g(x) \cdot_\star d_\star x \right)$$

$$= \quad a \cdot_\star \int_\star f(x) \cdot_\star d_\star x +_\star b \cdot_\star \int_\star g(x) \cdot_\star d_\star x.$$

This completes the proof.

2. We have

$$\int_\star (a \cdot_\star f(x) -_\star b \cdot_\star g(x)) \cdot_\star d_\star x = a \cdot_\star \int_\star f(x) \cdot_\star d_\star x -_\star b \cdot_\star \int_\star g(x) \cdot_\star d_\star x.$$

Proof Note that

$$a \cdot_\star f(x) -_\star b \cdot_\star g(x) \quad = \quad e^{\log a \log f(x)} -_\star e^{\log b \log g(x)}$$

$$= \quad e^{\log a \log f(x) - \log b \log g(x)}, \quad x \in \mathbb{R}_\star.$$

Hence,

$$\int_\star (a \cdot_\star f(x) -_\star b \cdot_\star g(x)) \cdot_\star d_\star x \quad = \quad \int_\star e^{\log a \log f(x) - \log b \log gx} \cdot_\star d_\star x$$

$$= \quad e^{\int \frac{1}{x} \log e^{\log a \log f(x) - \log b \log g(x)} dx}$$

$$= \quad e^{\int \frac{1}{x} \log a \log f(x) dx - \int \frac{1}{x} \log b \log g(x) dx}$$

$$= \quad e^{\log a \int \frac{1}{x} \log f(x) dx - \log b \int \frac{1}{x} \log g(x) dx}$$

$$= \quad \frac{e^{\log a \log e^{\int \frac{1}{x} \log f(x) dx}}}{e^{\log b \log e^{\int \frac{1}{x} \log g(x) dx}}}$$

$$= \quad \frac{\left(a \cdot_\star e^{\int \frac{1}{x} \log f(x) dx} \right)}{\left(b \cdot_\star e^{\int \frac{1}{x} \log g(x) dx} \right)}$$

$$= \quad \frac{(a \cdot_\star \int_\star f(x) \cdot_\star d_\star x)}{(b \cdot_\star \int_\star g(x) \cdot_\star d_\star x)}$$

$$= a \cdot_\star \int_\star f(x) \cdot_\star d_\star x -_\star b \cdot_\star \int_\star g(x) \cdot_\star d_\star x.$$

This completes the proof.

3. We have

$$\int_{\star a}^{a} f(x) \cdot_\star d_\star x = 0_\star.$$

Proof We have

$$
\begin{aligned}
\int_{\star a}^{a} f(x) \cdot_\star d_\star x &= e^{\int_{a}^{a} \frac{1}{x} \log f(x) dx} \\[2mm]
&= e^{0} \\[2mm]
&= 1 \\[2mm]
&= 0_\star.
\end{aligned}
$$

This completes the proof.

4. We have

$$\int_{\star a}^{b} f(x) \cdot_\star d_\star x = -_\star \int_{\star b}^{a} f(x) \cdot_\star d_\star x.$$

Proof We have

$$
\begin{aligned}
\int_{\star a}^{b} f(x) \cdot_\star d_\star x &= e^{\int_{a}^{b} \frac{1}{x} \log f(x) dx} \\[2mm]
&= e^{-\int_{b}^{a} \frac{1}{x} \log f(x) dx} \\[2mm]
&= -_\star e^{\int_{b}^{a} \frac{1}{x} \log f(x) dx} \\[2mm]
&= -_\star \int_{\star b}^{a} f(x) \cdot_\star d_\star x.
\end{aligned}
$$

This completes the proof.

5. We have

$$\int_{\star a}^{b} f(x) \cdot_\star d_\star x = \int_{\star a}^{c} f(x) \cdot_\star d_\star x +_\star \int_{\star c}^{b} f(x) \cdot_\star d_\star x.$$

Proof We have

$$\int_{\star a}^{b} f(x) \cdot_\star d_\star x = e^{\int_{a}^{b} \frac{1}{x} \log f(x) dx}$$

$$= e^{\int_a^c \frac{1}{x}\log f(x)dx + \int_c^b \frac{1}{x}\log f(x)dx}$$

$$= \left(e^{\int_a^c \frac{1}{x}\log f(x)dx}\right)\left(e^{\int_c^b \frac{1}{x}\log f(x)dx}\right)$$

$$= \left(\int_a^c f(x)\cdot_\star d_\star x\right)\left(\int_c^b f(x)\cdot_\star d_\star x\right)$$

$$= \int_{\star a}^c f(x)\cdot_\star d_\star x +_\star \int_{\star c}^b f(x)\cdot_\star d_\star x.$$

This completes the proof.

Example 3.4. We have

$$\int_\star (3\cdot_\star \sin_\star x -_\star 2\cdot_\star \cos_\star x)\cdot_\star d_\star x$$

$$= 3\cdot_\star \int_\star \sin_\star x\cdot_\star d_\star x -_\star 2\cdot_\star \int_\star \cos_\star x\cdot_\star d_\star x$$

$$= c(-_\star 3\cdot_\star \cos_\star x -_\star 2\cdot_\star \sin_\star x), \quad x\in\mathbb{R}_\star, \quad c>0.$$

Example 3.5. We have

$$\int_\star (2/_\star(\cos_\star x)^{2\star} -_\star 3)\cdot_\star d_\star x = \int_\star 2/_\star(\cos_\star x)^{2\star}\cdot_\star d_\star x -_\star \int_\star 3\cdot_\star d_\star x$$

$$= 2\cdot_\star \int_\star (1_\star/_\star(\cos_\star x)^{2\star})\cdot_\star d_\star x -_\star x^{\log 3}$$

$$= c\left(2\cdot_\star e^{\tan(\log x)} -_\star x^{\log 3}\right), \quad x\in\mathbb{R}_\star, \quad c>0.$$

Example 3.6. We have

$$\int_\star 4/_\star(1_\star -_\star x_\star^{2\star})^{\frac{1}{2}\star}\cdot_\star d_\star x = 4\cdot_\star \int_\star 1_\star/_\star(1_\star -_\star x_\star^{2\star})^{\frac{1}{2}\star}\cdot_\star d_\star x$$

$$= ce^{\arcsin(\log x)}, \quad x\in\mathbb{R}_\star, \quad c>0.$$

Exercise 3.2. Find

$$\int_\star (\cos_\star x +_\star (1_\star/_\star x_\star))\cdot_\star d_\star x.$$

Answer 3.2.

$$c(\sin_\star x +_\star |\log x|), \quad x\in\mathbb{R}_\star, \quad c>0.$$

3.4 Multiplicative Integration by Substitution

Suppose that $f, g : \mathbb{R}_\star \to \mathbb{R}_\star, f \in \mathscr{C}_\star(\mathbb{R}_\star), g \in \mathscr{C}_\star^1(\mathbb{R}_\star)$. Let also, $a, b, c, d \in \mathbb{R}_\star, g(a) = c$ and $g(b) = d$. Then

$$\int_{\star a}^b f(g(t)) \cdot_\star g^\star(t) \cdot_\star d_\star t \quad = \quad \int_{\star a}^b f(g(t)) \cdot_\star d_\star g(t)$$

$$= \quad \int_{\star c}^d f(s) \cdot_\star d_\star s.$$

Example 3.7. We will compute

$$I = \int_{\star 1}^2 \sin_\star (t^2 + 1) \cdot_\star e^{\frac{2t^2}{t^2+1}} \cdot_\star d_\star t.$$

Let

$$g(t) = t^2 + 1, \quad t \in \mathbb{R}_\star.$$

Then

$$g'(t) = 2t, \quad t \in \mathbb{R}_\star,$$

and

$$g^\star(t) \quad = \quad e^{t \frac{g'(t)}{g(t)}}$$

$$= \quad e^{\frac{2t^2}{t^2+1}}, \quad t \in \mathbb{R}_\star,$$

$$g(1) \quad = \quad 2,$$

$$g(2) \quad = \quad 5.$$

Hence,

$$I \quad = \quad \int_{\star 2}^5 \sin_\star s \cdot_\star d_\star s$$

$$= \quad e^{-\cos(\log s)} \Big|_{s=2}^{s=5}$$

$$= \quad e^{-\cos(\log 5)} -_\star e^{-\cos(\log 2)}$$

$$= \quad e^{\cos(\log 2) - \cos(\log 5)}.$$

Example 3.8. We will compute

$$I = \int_{\star 1}^{e^{\frac{\pi}{2}}} (\cos_\star t)^{4_\star} \cdot_\star \sin_\star t \cdot_\star d_\star t.$$

Let

$$g(t) = \cos_\star t, \quad t \in \mathbb{R}_\star.$$

Then

$$g^\star(t) \quad = \quad -_\star \sin_\star t,$$

$$g(1) \quad = \quad e^{\cos(\log 1)}$$

$$= \quad e^{\cos 0}$$

$$= \quad e,$$

$$g\left(e^{\frac{\pi}{2}}\right) \quad = \quad e^{\cos\left(\log e^{\frac{\pi}{2}}\right)}$$

$$= \quad e^{\cos\frac{\pi}{2}}$$

$$= \quad e^0$$

$$= \quad 1.$$

Thus,

$$I \quad = \quad -_\star \int_{\star 1}^{e^{\frac{\pi}{2}}} (\cos_\star t)^{4_\star} \cdot_\star d_\star \cos_\star t$$

$$= \quad -_\star \int_{\star e}^{1} x^{4_\star} \cdot_\star d_\star x$$

$$= \quad -_\star \left(e^{\frac{1}{5}(\log x)^5} \Big|_{x=e}^{x=1} \right)$$

$$= \quad -_\star \left(1 -_\star e^{\frac{1}{5}} \right)$$

$$= \quad -_\star \frac{1}{e^{\frac{1}{5}}}$$

$$= \quad e^{\frac{1}{5}}.$$

Example 3.9. We will compute

$$I = \int_{\star 1}^{e^{\frac{\pi}{2}}} \left(1_\star/_\star(\cos_\star(\sin_\star t))^{2_\star}\right) \cdot_\star \cos_\star t \cdot_\star d_\star t.$$

Let

$$g(t) = \sin_\star t, \quad t \in \mathbb{R}_\star.$$

Then

$$g^\star(t) = \cos_\star t, \quad t \in \mathbb{R}_\star,$$

and

$$
\begin{aligned}
g(1) &= e^{\sin(\log 1)} \\
&= e^{\sin 0} \\
&= e^0 \\
&= 1, \\
g\left(e^{\frac{\pi}{2}}\right) &= e^{\sin\left(\log e^{\frac{\pi}{2}}\right)} \\
&= e^{\sin\frac{\pi}{2}} \\
&= e.
\end{aligned}
$$

Thus,

$$
\begin{aligned}
I &= \int_{\star 1}^{e^{\frac{\pi}{2}}} \left(1_\star/_\star(\cos_\star(\sin_\star t))^{2_\star}\right) \cdot_\star d_\star \sin_\star t \\
&= \int_{\star 1}^{e} \left(1_\star/_\star(\cos_\star t)^{2_\star}\right) \cdot_\star d_\star t \\
&= \left. e^{\tan(\log t)} \right|_{t=1}^{t=e} \\
&= e^{\tan(\log e)} -_\star e^{\tan(\log 1)} \\
&= e^{\tan 1} -_\star 1 \\
&= e^{\tan 1}.
\end{aligned}
$$

Exercise 3.3. Compute

$$\int_{\star 1}^{2} e^{2} \cdot_{\star} (t^{2})^{5\star} \cdot_{\star} d_{\star}t.$$

Answer 3.3. $e^{\frac{32}{3}(\log 2)^{6}}$.

3.5   Multiplicative Integration by Parts

Theorem 3.1. *Let $f, g : \mathbb{R}_{\star} \to \mathbb{R}_{\star}$, $f, g \in \mathscr{C}_{\star}^{1}(\mathbb{R}_{\star})$. Let also, $a, b \in \mathbb{R}_{\star}$, $a < b$. Then*

$$\int_{\star a}^{b} f^{\star}(x) \cdot_{\star} g(x) \cdot_{\star} d_{\star}x \;=\; f(x) \cdot_{\star} g(x) \Big|_{x=a}^{x=b}$$

$$-_{\star} \int_{\star a}^{b} f(x) \cdot_{\star} g^{\star}(x) \cdot_{\star} d_{\star}x.$$

Proof   We have

$$(f \cdot_{\star} g)^{\star}(x) = f^{\star}(x) \cdot_{\star} g(x) +_{\star} f(x) \cdot_{\star} g^{\star}(x), \quad x \in [a, b].$$

Hence,

$$f^{\star}(x) \cdot_{\star} g(x) = (f \cdot_{\star} g)^{\star}(x) -_{\star} f(x) \cdot_{\star} g^{\star}(x), \quad x \in [a, b],$$

and

$$\int_{\star a}^{b} f^{\star}(x) \cdot_{\star} g(x) \cdot_{\star} d_{\star}x \;=\; \int_{\star a}^{b} ((f \cdot_{\star} g)^{\star}(x) -_{\star} f(x) \cdot_{\star} g^{\star}(x)) \cdot_{\star} d_{\star}x$$

$$=\; \int_{\star a}^{b} (f \cdot_{\star} g)^{\star}(x) \cdot_{\star} d_{\star}x -_{\star} \int_{\star a}^{b} f(x) \cdot_{\star} g^{\star}(x) \cdot_{\star} d_{\star}x$$

$$=\; (f \cdot_{\star} g)(x) \Big|_{x=a}^{x=b} -_{\star} \int_{\star a}^{b} f(x) \cdot_{\star} g^{\star}(x) \cdot_{\star} d_{\star}x.$$

This completes the proof.

Example 3.10.   We will compute

$$I = \int_{\star 1}^{e} x^{3} \cdot_{\star} \cos_{\star} x \cdot_{\star} d_{\star}x.$$

We have

$$I \;=\; \int_{\star 1}^{e} x^{3} \cdot_{\star} \sin_{\star}^{\star} x \cdot_{\star} d_{\star}x$$

$$= \quad x^3 \cdot_\star \sin_\star x \Big|_{x=1}^{x=e}$$

$$-_\star \int_{\star 1}^e (x^3)^\star \cdot_\star \sin_\star x \cdot_\star d_\star x$$

$$= \quad e^3 \cdot_\star e^{\sin(\log e)} -_\star 1 \cdot_\star e^{\sin(\log 1)}$$

$$-_\star e^3 \cdot_\star \int_{\star 1}^e \sin_\star x \cdot_\star d_\star x$$

$$= \quad e^3 \cdot_\star e^{\sin 1} -_\star 1 \cdot_\star 1$$

$$+_\star e^3 \cdot_\star \cos_\star x \Big|_{x=1}^{x=e}$$

$$= \quad e^{3\sin 1} -_\star 1 +_\star e^3 \cdot_\star (\cos_\star e -_\star \cos_\star 1)$$

$$= \quad e^{3\sin 1} +_\star e^3 \cdot_\star \left(e^{\cos(\log e)} -_\star e^{\cos(\log 1)} \right)$$

$$= \quad e^{3\sin 1} +_\star e^3 \cdot_\star \left(e^{\cos 1} -_\star e^1 \right)$$

$$= \quad e^{3\sin 1} +_\star e^3 \cdot_\star e^{\cos 1 - 1}$$

$$= \quad e^{3\sin 1} +_\star e^{3(\cos 1 - 1)}$$

$$= \quad e^{3(\sin 1 + \cos 1 - 1)}.$$

Example 3.11. Now, we will compute

$$I = \int_{\star 1}^e e^{x \log 2} \cdot_\star x^2 \cdot_\star d_\star x.$$

We have

$$(2^x)^\star = e^{x \log 2}, \quad x \in \mathbb{R}_\star.$$

Then

$$I = \int_{\star 1}^e (2^x)^\star \cdot_\star x^2 \cdot_\star d_\star x$$

$$= \quad 2^x \cdot_\star x^{2\star} \Big|_{x=1}^{x=e} -_\star \int_{\star 1}^e 2^x \cdot_\star e^2 \cdot_\star d_\star x$$

$$= \quad 2^e \cdot_\star e^{(\log e)^2} -_\star e^2 \cdot_\star \int_{\star 1}^e 2^x \cdot_\star d_\star x$$

$$= \quad 2^e \cdot_\star e -_\star e^2 \cdot_\star e^{\log 2(e-1)}$$

$$= \quad e^{e\log 2} -_\star e^{2\log 2(e-1)}$$

$$= \quad \frac{e^{e\log 2}}{e^{2(e-1)\log 2}}$$

$$= \quad e^{(2-e)\log 2}.$$

Example 3.12. Now, we will compute

$$I = \int_{\star e^2}^{e^3} e^x \cdot_\star \cos_\star x \cdot_\star d_\star x.$$

We have

$$I \quad = \quad \int_{\star e^2}^{e^3} \cos_\star x \cdot_\star d_\star e^x$$

$$= \quad \cos_\star x \cdot_\star e^x \Big|_{x=e^2}^{x=e^3} +_\star \int_{\star e^2}^{e^3} \sin_\star x \cdot_\star e^x \cdot_\star d_\star x$$

$$= \quad e^{x\cos(\log x)} \Big|_{x=e^2}^{x=e^3} +_\star \int_{\star e^2}^{e^3} \sin_\star x \cdot_\star d_\star e^x$$

$$= \quad e^{e^3\cos 3} -_\star e^{e^2\cos 2} +_\star \sin_\star x \cdot_\star e^x \Big|_{x=e^2}^{x=e^3}$$

$$\quad -_\star \int_{\star e^2}^{e^3} \cos_\star x \cdot_\star e^x \cdot_\star d_\star x$$

$$= \quad e^{e^3\cos 3 - e^2\cos 2} +_\star e^{x\sin(\log x)} \Big|_{x=e^2}^{x=e^3} -_\star I$$

$$= \quad e^{e^3\cos 3 - e^2\cos 2} +_\star e^{e^3\sin 3} -_\star e^{e^2\sin 2} -_\star I$$

$$= \quad e^{e^3\cos 3 - e^2\cos 2} +_\star e^{e^3\sin 3 - e^2\sin 2} -_\star I$$

$$= \quad e^{e^3(\cos 3 + \sin 3) - e^2(\cos 2 + \sin 2)} -_\star I,$$

whereupon

$$e^2 \cdot_\star I = e^{e^3(\cos 3 + \sin 3) - e^2(\cos 2 + \sin 2)}$$

and

$$I = ee^{3}(\cos 3+\sin 3)-e^{2}(\cos 2+\sin 2)/_{\star}e^{2}$$

$$= e^{\frac{1}{2}\left(e^{3}(\cos 3+\sin 3)-e^{2}(\cos 2+\sin 2)\right)}.$$

Exercise 3.4. Find

$$\int_{\star e^{2}}^{e^{4}} e^{x} \cdot_{\star} \sin_{\star} x \cdot_{\star} d_{\star}x.$$

Answer 3.4.

$$e^{\frac{1}{2}\left(e^{2}(\cos 2-\sin 2)-e^{4}(\cos 4-\sin 4)\right)}.$$

3.6 Inequalities for Multiplicative Integrals

In this section, we will deduct some important inequalities for multiplicative integrals. Let $a, b \in \mathbb{R}_{\star}$, $a < b$.

Theorem 3.2. *Let* $f \in \mathscr{C}_{\star}([a,b])$ *and* $f \geq 0_{\star}$ *on* $[a,b]$. *Then*

$$\int_{\star a}^{b} f(x) \cdot_{\star} d_{\star}x \geq 0_{\star}.$$

Proof We have

$$f(x) \geq 0_{\star}$$

$$= 1, \quad x \in [a,b].$$

Hence,

$$\log f(x) \geq 0, \quad x \in [a,b],$$

and

$$\int_{\star a}^{b} f(x) \cdot_{\star} d_{\star}x = e^{\int_{a}^{b} \frac{1}{x} \log f(x)dx}$$

$$\geq e^{0}$$

$$= 1$$

$$= 0_{\star}.$$

This completes the proof.

Theorem 3.3. *Let $f, g \in \mathscr{C}_\star([a,b])$ and $0 \leq f(x) \leq g(x)$, $x \in [a,b]$. Then*

$$\int_{\star a}^{b} g(x) \cdot_\star d_\star x \geq \int_{\star a}^{b} f(x) \cdot_\star d_\star x.$$

Proof We have that

$$1 = 0_\star$$

$$\leq \frac{g(x)}{f(x)}$$

$$= g(x) -_\star f(x), \quad x \in [a,b].$$

Hence, applying Theorem 3.2, we find

$$0_\star \leq \int_{\star a}^{b} (g(x) -_\star f(x)) \cdot_\star d_\star x$$

$$= \int_{\star a}^{b} g(x) \cdot_\star d_\star x -_\star \int_{\star a}^{b} f(x) \cdot_\star d_\star x,$$

whereupon

$$\int_{\star a}^{b} f(x) \cdot_\star d_\star x \leq \int_{\star a}^{b} g(x) \cdot_\star d_\star x.$$

This completes the proof.

Theorem 3.4. *Let $f \in \mathscr{C}_\star(\mathbb{R}_\star)$, $f : \mathbb{R}_\star \to \mathbb{R}_\star$. Then*

$$\left| \int_{\star a}^{b} f(x) \cdot_\star d_\star x \right|_\star \leq \int_{\star a}^{b} |f(x)|_\star \cdot_\star d_\star x \leq M \cdot_\star (b -_\star a),$$

where $M = \sup\limits_{x \in [a,b]} |f(x)|_\star$.

Proof Note that

$$-_\star |f(x)|_\star \leq f(x) \leq |f(x)|_\star, \quad x \in [a,b],$$

and hence,

$$-_\star \int_{\star a}^{b} |f(x)|_\star \cdot_\star d_\star x \leq \int_{\star a}^{b} f(x) \cdot_\star d_\star x \leq \int_{\star a}^{b} |f(x)|_\star \cdot_\star d_\star x.$$

Therefore,

$$\left| \int_{\star a}^{b} f(x) \cdot_\star d_\star x \right|_\star \leq \int_{\star a}^{b} |f(x)|_\star \cdot_\star d_\star x$$

$$\leq M \cdot_\star \int_{\star a}^b d_\star x$$

$$= M \cdot_\star (b -_\star a).$$

This completes the proof.

Theorem 3.5. *Let* $f, g \in \mathscr{C}_\star([a,b])$, $f, g : \mathbb{R}_\star \to \mathbb{R}_\star$,

$$m = \inf_{x \in [a,b]} |g(x)|_\star, \quad M = \sup_{x \in [a,b]} |g(x)|_\star. \tag{3.1}$$

Then

$$m \cdot_\star \int_{\star a}^b f(x) \cdot_\star d_\star x \leq \int_{\star a}^b f(x) \cdot_\star g(x) \cdot_\star d_\star x \leq M \cdot_\star \int_{\star a}^b f(x) \cdot_\star d_\star x.$$

Proof We have

$$m \cdot_\star f(x) \leq f(x) \cdot_\star g(x) \leq M \cdot_\star f(x), \quad x \in [a,b],$$

and hence,

$$\int_{\star a}^b m \cdot_\star f(x) \cdot_\star d_\star x = m \cdot_\star \int_{\star a}^b f(x) \cdot_\star d_\star x$$

$$\leq \int_{\star a}^b f(x) \cdot_\star g(x) \cdot_\star d_\star x$$

$$\leq \int_{\star a}^b M \cdot_\star f(x) \cdot_\star d_\star x$$

$$= M \cdot_\star \int_{\star a}^b f(x) \cdot_\star d_\star x.$$

This completes the proof.

Theorem 3.6 (The Multiplicative Cauchy-Schwartz Inequality). *Let* $f, g \in \mathscr{C}_\star(\mathbb{R}_\star)$, $f, g : \mathbb{R}_\star \to \mathbb{R}_\star$. *Then*

$$\left(\int_{\star a}^b f(x) \cdot_\star g(x) \cdot_\star d_\star x \right)^{2_\star} \leq \left(\int_{\star a}^b (f(x))^{2_\star} \cdot_\star d_\star x \right) \cdot_\star \left(\int_{\star a}^b (g(x))^{2_\star} \cdot_\star d_\star x \right).$$

Proof Applying the classical Cauchy-Schwartz inequality, we get

$$\left(\int_{\star a}^b f(x) \cdot_\star g(x) \cdot_\star d_\star x \right)^{2_\star} = e^{\left(\log \int_{\star a}^b f(x) \cdot_\star g(x) \cdot_\star d_\star x \right)^2}$$

$$= e^{\left(\log \int_{\star a}^{b} e^{\log f(x)\log g(x)} \cdot_{\star} d_{\star}x\right)^2}$$

$$= e^{\left(\log e^{\int_{a}^{b} \frac{1}{x}\log e^{\log f(x)\log g(x)} dx}\right)^2}$$

$$= e^{\left(\int_{a}^{b} \frac{1}{x}\log f(x)\log g(x) dx\right)^2}$$

$$= e^{\left(\int_{a}^{b} \left(\frac{1}{\sqrt{x}}\log f(x)\right)\left(\frac{1}{\sqrt{x}}\log g(x)\right) dx\right)^2}$$

$$\leq e^{\left(\int_{a}^{b} \frac{1}{x}(\log f(x))^2 dx\right)\left(\int_{a}^{b} \frac{1}{x}(\log g(x))^2 dx\right)}$$

$$= e^{\int_{a}^{b} \frac{1}{x}(\log f(x))^2 dx} \cdot_{\star} e^{\int_{a}^{b} \frac{1}{x}(\log g(x))^2 dx}.$$

Note that

$$\log(f(x))^{2\star} = \log e^{(\log f(x))^2}$$

$$= (\log f(x))^2, \quad x \in [a,b].$$

As above,

$$\log(g(x))^{2\star} = (\log g(x))^2, \quad x \in [a,b].$$

Therefore,

$$\left(\int_{\star a}^{b} f(x) \cdot_{\star} g(x) \cdot_{\star} d_{\star}x\right)^{2\star} \leq e^{\int_{a}^{b} \frac{1}{x}(\log f(x))^2 dx} \cdot_{\star} e^{\int_{a}^{b} \frac{1}{x}(\log g(x))^2 dx}$$

$$= e^{\int_{a}^{b} \frac{1}{x}\log(f(x))^{2\star} dx} \cdot_{\star} e^{\int_{a}^{b} \frac{1}{x}\log(g(x))^{2\star} dx}$$

$$= \left(\int_{\star a}^{b} (f(x))^{2\star} \cdot_{\star} d_{\star}x\right) \cdot_{\star} \left(\int_{\star a}^{b} (g(x))^{2\star} \cdot_{\star} d_{\star}x\right).$$

This completes the proof.

3.7 Mean Value Theorems for Multiplicative Integrals

Suppose that $a, b \in \mathbb{R}_{\star}$, $a < b$.

Theorem 3.7. *Let* $f, g \in \mathscr{C}_{\star}([a,b])$, $f, g : \mathbb{R}_{\star} \to \mathbb{R}_{\star}$, $g(x) \geq 0_{\star}$, $x \in [a,b]$. *Let also,*

$$m = \inf_{x \in [a,b]} f(x), \quad M = \sup_{x \in [a,b]} f(x). \tag{3.2}$$

Then there is a $\mu \in [m, M]$ so that

$$\int_{\star a}^{b} f(x) \cdot_{\star} g(x) \cdot_{\star} d_{\star} x = \mu \cdot_{\star} \int_{\star a}^{b} g(x) \cdot_{\star} d_{\star} x.$$

Proof If

$$\int_{\star a}^{b} g(x) \cdot_{\star} d_{\star} x = 0_{\star},$$

then $g(x) = 0_{\star}$, $x \in [a, b]$, and

$$\int_{\star a}^{b} f(x) \cdot_{\star} g(x) \cdot_{\star} d_{\star} x = 0_{\star},$$

and the assertion is proved. Let

$$\int_{\star a}^{b} g(x) \cdot_{\star} d_{\star} x > 0_{\star}.$$

Then, applying Theorem 3.5, it follows that

$$m \cdot_{\star} \int_{\star a}^{b} g(x) \cdot_{\star} d_{\star} x \quad \leq \quad \int_{\star a}^{b} f(x) \cdot_{\star} g(x) \cdot_{\star} d_{\star} x$$

$$\leq \quad M \cdot_{\star} \int_{\star a}^{b} g(x) \cdot_{\star} d_{\star} x.$$

Hence,

$$m \quad \leq \quad \int_{\star a}^{b} f(x) \cdot_{\star} g(x) \cdot_{\star} d_{\star} x /_{\star} \int_{\star a}^{b} g(x) \cdot_{\star} d_{\star} x$$

$$\leq \quad M.$$

Set

$$\mu = \int_{\star a}^{b} f(x) \cdot_{\star} g(x) \cdot_{\star} d_{\star} x /_{\star} \int_{\star a}^{b} g(x) \cdot_{\star} d_{\star} x,$$

whereupon

$$\int_{\star a}^{b} f(x) \cdot_{\star} g(x) \cdot_{\star} d_{\star} x = \mu \cdot_{\star} \int_{\star a}^{b} g(x) \cdot_{\star} d_{\star} x$$

This completes the proof.

Corollary 3.1. Suppose that all conditions of Theorem 3.7 hold. Then there is a $\mu \in [m, M]$ so that

$$\int_{\star a}^{b} f(x) \cdot_{\star} d_{\star} x = \mu \cdot_{\star} (b -_{\star} a).$$

Proof We apply Theorem 3.7 for $g(x) = 1_{\star}$ and get the desired result. This completes the proof.

Theorem 3.8. *Let* $f, g \in \mathscr{C}_*[a,b], f, g : [a,b] \to \mathbb{R}_*, g(x) > 0_*, x \in [a,b]$. *Then there is a* $c \in [a,b]$ *so that*

$$\int_{*a}^{b} f(x) \cdot_* g(x) \cdot_* d_*x = f(c) \cdot_* \int_{*a}^{b} g(x) \cdot_* d_*x.$$

Proof Let

$$m = \inf_{x \in [a,b]} f(x), \quad M = \sup_{x \in [a,b]} f(x).$$

Then there are $x_1, x_2 \in [a,b]$ so that

$$m = f(x_1), \quad M = f(x_2).$$

By Theorem 3.7, it follows that there is a $\mu \in [m, M]$ so that

$$\int_{*a}^{b} f(x) \cdot_* g(x) \cdot_* d_*x = \mu \cdot_* \int_{*a}^{b} g(x) \cdot_* d_*x.$$

Next, there is a $c \in [x_1, x_2]$ so that

$$\mu = f(c).$$

Hence, we get the desired result. This completes the proof.

3.8 Advanced Practical Problems

Problem 3.1. Let $f(x) = e^{x^2 \cos x}$, $x \in \mathbb{R}_*$. Find

$$\int_{*\frac{\pi}{3}}^{\frac{\pi}{2}} f(x) d_*.$$

Answer 3.5.

$$e^{\frac{\pi}{2} - \frac{\pi\sqrt{3}}{6} - \frac{1}{2}}.$$

Problem 3.2. Find

$$\int_* (1_*/_* x +_* \sinh_* x) \cdot_* d_*x.$$

Answer 3.6.

$$c(|\log x| +_* \cosh_* x), \quad x \in \mathbb{R}_*, \quad c > 0.$$

Problem 3.3. Compute

$$\int_{*1}^{2} e^3 \cdot_* t^{2_*} \cdot_* d_*t.$$

Answer 3.7. $e^{(\log 2)^3}$.

Problem 3.4. Compute

$$\int_{\star e^3}^{e^7} e^x \cdot_\star \cos_\star x \cdot_\star d_\star x.$$

Answer 3.8.

$$e^{\frac{1}{2}\left(e^7(\cos 7 + \sin 7) - e^3(\cos 3 + \sin 3)\right)}.$$

Problem 3.5. Compute

$$\int_{\star 1}^{e} x^{3\star} \cdot_\star x^{4\star} \cdot_\star x^{5\star} \cdot_\star d_\star x.$$

Answer 3.9. $e^{\frac{1}{13}}$.

4

Improper Multiplicative Integrals

In this chapter, we introduce the improper multiplicative integrals on finite and infinite intervals. We give and prove some criteria for convergence and divergence of improper multiplicative integrals.

4.1 Definition for Improper Multiplicative Integrals over Finite Intervals

Suppose that $a, b \in \mathbb{R}_\star$, $a < b$.

Definition 4.1. Let $f : [a, b) \to \mathbb{R}_\star$, f be multiplicative integrable over $[a, \eta]$ for any $\eta \in [a, b)$ and f be eventually unbounded in a neighborhood of b. If the limit

$$\lim_{\eta \to b-0} \int_{\star a}^{\eta} f(x) \cdot_\star d_\star x$$

exists, then the improper multiplicative integral

$$\int_{\star a}^{b} f(x) \cdot_\star d_\star x$$

is said to be convergent. Otherwise, it is called divergent improper multiplicative integral.

Definition 4.2. Let $f : (a, b] \to \mathbb{R}_\star$, f be multiplicative integrable over $[\xi, b]$ for any $\xi \in (a, b]$ and f be eventually unbounded in a neighborhood of a. If the limit

$$\lim_{\xi \to a+0} \int_{\star \xi}^{b} f(x) \cdot_\star d_\star x$$

exists, then the improper multiplicative integral

$$\int_{\star a}^{b} f(x) \cdot_\star d_\star x$$

is said to be convergent. Otherwise, it is called divergent improper multiplicative integral.

DOI: 10.1201/9781003299080-4

Definition 4.3. Let $f : \mathbb{R}_\star \to \mathbb{R}_\star$ be multiplicative integrable over $[a, \xi]$ and $[\eta, b]$ for any $\xi \in [a, c)$ and any $\eta \in (c, b]$, and f be eventually unbounded in a neighborhood of the point c, then we define

$$\int_{\star a}^b f(x) \cdot_\star d_\star x = \int_{\star a}^c f(x) \cdot_\star d_\star x +_\star \int_{\star c}^b f(x) \cdot_\star d_\star x.$$

Example 4.1. Let $f : (2,3] \to \mathbb{R}_\star$ is defined by

$$f(x) = e^{\frac{x}{\sqrt{x-2}}}, \quad x \in (2,3].$$

Then

$$
\begin{aligned}
\lim_{\xi \to 2+0} \int_{\star\xi}^3 f(x) \cdot_\star d_\star x &= \lim_{\xi \to 2+0} e^{\int_\xi^3 \frac{1}{x} \log e^{\frac{x}{\sqrt{x-2}}} dx} \\
&= \lim_{\xi \to 2+0} e^{\int_\xi^3 \frac{1}{\sqrt{x-2}} dx} \\
&= \lim_{\xi \to 2+0} e^{2\sqrt{x-2}\Big|_{x=\xi}^{x=3}} \\
&= e^2.
\end{aligned}
$$

Thus, the considered improper multiplicative integral is convergent.

Example 4.2. Let

$$f(x) = e^{\frac{x^2}{\sqrt{16-x^2}}}, \quad x \in [2,4).$$

We have that f is unbounded in a neighborhood of $x = 4$. Moreover,

$$
\begin{aligned}
\lim_{\eta \to 4} \int_2^\eta f(x) \cdot_\star d_\star x &= \lim_{\eta \to 4} e^{\int_2^\eta \frac{1}{x} \log e^{\frac{x^2}{\sqrt{16-x^2}}} dx} \\
&= \lim_{\eta \to 4} e^{\int_2^\eta \frac{x}{\sqrt{16-x^2}} dx} \\
&= \lim_{\eta \to 4} e^{-\frac{1}{2} \int_2^\eta \frac{d(16-x^2)}{\sqrt{16-x^2}}} \\
&= \lim_{\eta \to 4} e^{-\frac{1}{2} \frac{\sqrt{16-x^2}}{\frac{1}{2}}\Big|_{x=2}^{x=\eta}} \\
&= \lim_{\eta \to 4} e^{-\sqrt{16-x^2}\Big|_{x=2}^{x=\eta}}
\end{aligned}
$$

$$= e^{\sqrt{12}}$$

$$= e^{2\sqrt{3}}.$$

Therefore, the considered improper integral is convergent.

Example 4.3. Let

$$f(x) = e^{\frac{x}{(x-3)^{\frac{1}{3}}}}, \quad x \in [2,4].$$

Note that f is unbounded in a neighborhood of $x = 3$. We have

$$\int_{\star 2}^{4} f(x) \cdot_\star d_\star x = e^{\int_2^4 \frac{1}{x} \log e^{\frac{x}{(x-3)^{\frac{1}{3}}}} dx}$$

$$= \lim_{\eta \to 3} e^{\int_2^\eta \frac{dx}{(x-3)^{\frac{1}{3}}} + \int_\eta^4 \frac{dx}{(x-3)^{\frac{1}{3}}}}$$

$$= \lim_{\eta \to 3} e^{\frac{3}{2}(x-3)^{\frac{2}{3}} \Big|_{x=2}^{x=\eta} + \frac{3}{2}(x-3)^{\frac{2}{3}} \Big|_{x=\eta}^{x=4}}$$

$$= e^{\frac{3}{2} + \frac{3}{2}}$$

$$= e^3.$$

Thus, the considered improper multiplicative integral is convergent.

Exercise 4.1. Prove that

$$\int_{\star 2}^{7} e^{\frac{x}{(x-2)^{\frac{1}{4}}}} \cdot_\star d_\star x$$

is convergent.

4.2 Definition for Improper Multiplicative Integrals over Infinite Intervals

Suppose that $a \in \mathbb{R}_\star$.

Definition 4.4. Let $f : [a, \infty) \to \mathbb{R}_\star$, f be multiplicative integrable over $[a, \eta]$ for any $\eta \in [a, b]$. If the limit

$$\lim_{\eta \to \infty} \int_{\star a}^{\eta} f(x) \cdot_\star d_\star x$$

exists, then the improper multiplicative integral

$$\int_{\star a}^{\infty} f(x) \cdot_{\star} d_{\star}x$$

is said to be convergent. Otherwise, it is called divergent improper multiplicative integral.

Example 4.4. Let

$$f(x) = x, \quad x \in [2, \infty).$$

Then

$$\int_{\star 2}^{\infty} f(x) \cdot_{\star} d_{\star}x = \lim_{\eta \to \infty} \int_{\star 2}^{\eta} f(x) \cdot_{\star} d_{\star}x$$

$$= \lim_{\eta \to \infty} e^{\int_2^{\eta} \frac{1}{x} \log f(x) dx}$$

$$= \lim_{\eta \to \infty} e^{\int_2^{\eta} \frac{\log x}{x} dx}$$

$$= \lim_{\eta \to \infty} e^{\int_2^{\eta} \log x \, d \log x}$$

$$= \lim_{\eta \to \infty} e^{\left. \frac{1}{2}(\log x)^2 \right|_{x=2}^{\eta}}$$

$$= \infty.$$

Thus, the considered improper multiplicative integral is divergent.

Example 4.5. Let

$$f(x) = e^{\frac{1}{(\log x)^2}}, \quad x \in [2, \infty).$$

We have

$$\int_{\star 2}^{\infty} f(x) \cdot_{\star} d_{\star}x = \lim_{\eta \to \infty} \int_{\star 2}^{\eta} f(x) \cdot_{\star} d_{\star}x$$

$$= \lim_{\eta \to \infty} e^{\int_2^{\eta} \frac{1}{x} \log e^{\frac{1}{(\log x)^2}} dx}$$

$$= \lim_{\eta \to \infty} e^{\int_2^{\eta} \frac{1}{x(\log x)^2} dx}$$

$$= \lim_{\eta \to \infty} e^{\int_2^{\eta} \frac{d \log x}{(\log x)^2}}$$

$$= \lim_{\eta \to \infty} e^{-\frac{1}{\log x}} \Big|_{x=2}^{x=\eta}$$

$$= e^{\frac{1}{\log 2}}.$$

Therefore, the considered improper multiplicative integral is convergent.

Example 4.6. Let

$$f(x) = e^{(\log x)^3}, \quad x \in [e, \infty).$$

Then

$$\int_{*e}^{\infty} f(x) \cdot_* d_* x = \lim_{\eta \to \infty} \int_{*e}^{\eta} f(x) \cdot_* d_* x$$

$$= \lim_{\eta \to \infty} e^{\int_{e}^{\eta} \frac{1}{x} \log f(x) dx}$$

$$= \lim_{\eta \to \infty} e^{\int_{e}^{\eta} \frac{1}{x} (\log x)^3 dx}$$

$$= \lim_{\eta \to \infty} e^{\frac{1}{4}(\log x)^4} \Big|_{x=e}^{x=\eta}$$

$$= \infty.$$

Thus, the considered improper multiplicative integral is divergent.

Exercise 4.2. Prove that

$$\int_{*3}^{\infty} e^{\frac{1}{x^4}} \cdot_* d_* x$$

is convergent.

4.3 Properties of the Improper Multiplicative Integrals

Let $a, b \in \mathbb{R}_*$, $a < b$. With the symbol

$$\int_{*a}^{b} f(x) \cdot_* d_* x \qquad (4.1)$$

we will denote an improper multiplicative integral if

1. f is multiplicative integrable on $[a, \eta]$ for any $\eta \in [a, b)$ and it is eventually unbounded in a neighborhood of b, or

2. f is multiplicative integrable on $[a, \eta]$ for any $\eta \in [a, \infty)$.

Theorem 4.1. *Let the multiplicative integral* (4.1) *has a unique singularity at* $x = b$. *Then it is convergent if and only if for any* $\varepsilon > 0$ *there is a* $b_0 \in (a, b)$ *so that for any* $b_1, b_2 \in (b_0, b)$, $b_1 < b_2 < b$, *we have*

$$\left| \int_{\star b_1}^{b_2} f(x) \cdot_\star d_\star x \right|_\star < \varepsilon.$$

Proof The multiplicative integral (4.1) is convergent if the limit

$$\lim_{\eta \to b - 0} F(\eta) = \lim_{\eta \to b - 0} \int_{\star a}^{\eta} f(x) \cdot_\star d_\star x$$

exists. This limit exists if and only if there is a $b_0 \in (a, b)$ so that for any $b_1, b_2 \in (b_0, b)$, $b_1 < b_2$, we have

$$|F(b_2) -_\star F(b_1)|_\star < \varepsilon.$$

This completes the proof.

Theorem 4.2. *Let* $a < c < b$ *and the multiplicative integral* (4.1) *has a unique singularity at* $x = b$. *Then* (4.1) *is convergent if and only if the multiplicative integral*

$$\int_{\star c}^{b} f(x) \cdot_\star d_\star x \qquad (4.2)$$

is convergent.

Proof By Theorem 4.1, it follows that the multiplicative integral (4.2) is convergent if and only if for any $\varepsilon > 0$ there is a $b_3 \in (c, b)$ so that for any $b_1, b_2 \in (b_3, b)$, $b_1 < b_2$, we have

$$\left| \int_{\star b_1}^{b_2} f(x) \cdot_\star d_\star x \right|_\star < \varepsilon.$$

Now, we take $b_0 = b_3$ and we get the desired result. This completes the proof.

Theorem 4.3. *Suppose that* $f, g : \mathbb{R}_\star \to \mathbb{R}_\star$, $c, d \in \mathbb{R}_\star$, *and the multiplicative integrals*

$$\int_{\star a}^{b} (c \cdot_\star f(x) +_\star d \cdot_\star g(x)) \cdot_\star d_\star x,$$

and

$$\int_{\star a}^{b} f(x) \cdot_\star d_\star x, \quad \int_{\star a}^{b} g(x) \cdot_\star d_\star x$$

have unique singularities at $x = b$. *Then*

$$\int_{\star a}^{b} (c \cdot_\star f(x) +_\star d \cdot_\star g(x)) \cdot_\star d_\star x = c \cdot_\star \int_{\star a}^{b} f(x) \cdot_\star d_\star x +_\star d \cdot_\star \int_{\star a}^{b} g(x) \cdot_\star d_\star x.$$

Proof We have

$$\int_{\star a}^{b} (c \cdot_\star f(x) +_\star d \cdot_\star g(x)) \cdot_\star d_\star x = \lim_{\eta \to b - 0} \int_{\star a}^{\eta} (c \cdot_\star f(x) +_\star d \cdot_\star g(x)) \cdot_\star d_\star x$$

$$= \lim_{\eta \to b-0} \left(\int_{*a}^{\eta} c \cdot_* f(x) \cdot_* d_* x +_* \int_{*a}^{\eta} d \cdot_* g(x) \cdot_* d_* x \right)$$

$$= \lim_{\eta \to b-0} \int_{*a}^{\eta} c \cdot_* f(x) \cdot_* d_* x +_* \lim_{\eta \to b-0} \int_{*a}^{\eta} d \cdot_* g(x) \cdot_* d_* x$$

$$= c \cdot_* \lim_{\eta \to b-0} \int_{*a}^{\eta} f(x) \cdot_* d_* x +_* d \cdot_* \lim_{\eta \to b-0} \int_{*a}^{\eta} g(x) \cdot_* d_* x$$

$$= c \cdot_* \int_{*a}^{b} f(x) \cdot_* d_* x +_* d \cdot_* \int_{*a}^{b} g(x) \cdot_* d_* x.$$

This completes the proof.

Definition 4.5. The multiplicative integral (4.1) is said to be absolutely convergent if the multiplicative integral

$$\int_{*a}^{b} |f(x)|_* \cdot_* d_* x \tag{4.3}$$

is convergent.

Theorem 4.4. *Suppose that the multiplicative integral* (4.1) *has a unique singularity at* $x = b$. *If the multiplicative integral* (4.1) *is absolutely convergent, then it is convergent.*

Proof Since the multiplicative integral (4.3) is convergent, we have

$$\lim_{\eta \to b-0} \int_{*a}^{\eta} |f(x)|_* \cdot_* d_* x < \infty.$$

Hence,

$$\lim_{\eta \to b-0} \left| \int_{*a}^{\eta} f(x) \cdot_* d_* x \right|_* \leq \lim_{\eta \to b-0} \int_{*a}^{\eta} |f(x)|_* \cdot_* d_* x$$

$$< \infty.$$

This completes the proof.

Example 4.7. Consider

$$I = \int_{*a}^{b} 1_* /_* (b -_* x)^{\alpha_*} \cdot_* d_* x.$$

We have

$$I = \int_{*a}^{b} 1_* /_* (b -_* x)^{\alpha_*} \cdot_* d_* x$$

$$= \int_{*a}^{b} e /_* e^{\left(\log \frac{b}{x} \right)^{\alpha}} \cdot_* d_* x$$

$$= \int_{\star a}^{b} e^{\overline{\left(\log \frac{b}{x}\right)^{\alpha}}} \cdot_{\star} d_{\star}x$$

$$= e^{\int_{a}^{b} \frac{1}{x} \log e^{\frac{1}{(\log b - \log x)^{\alpha}}} dx}$$

$$= e^{\int_{a}^{b} \frac{1}{x(\log b - \log x)^{\alpha}} dx}$$

$$= e^{\int_{a}^{b} \frac{1}{(\log b - \log x)^{\alpha}} d\log x}$$

$$= \begin{cases} \infty & \text{if} \quad \alpha \geq 1 \\[2ex] e^{\frac{(\log b - \log a)^{1-\alpha}}{1-\alpha}} & \text{if} \quad \alpha < 1. \end{cases}$$

Exercise 4.3. Prove that

$$\int_{\star 2}^{4} 1_{\star}/_{\star}(4 -_{\star} x)^{3\star} \cdot_{\star} d_{\star}x$$

is divergent.

4.4   Criteria for Comparison of Improper Multiplicative Integrals

Theorem 4.5. *Let the improper multiplicative integrals*

$$\int_{\star a}^{b} f(x) \cdot_{\star} d_{\star}x \tag{4.4}$$

and

$$\int_{\star a}^{b} g(x) \cdot_{\star} d_{\star}x \tag{4.5}$$

have a unique singularity at $x = b$. If

$$0_{\star} \leq f(x) \leq g(x), \quad x \in [a, b],$$

then by the convergence of the improper multiplicative integral (4.5) it follows the convergence of the improper multiplicative integral (4.4), and by the divergence of the improper multiplicative integral (4.4) it follows the divergence of the improper multiplicative integral (4.5).

Proof Note that for any $\eta \in [a, b]$ we have

$$0_{\star} \leq f(x) \leq g(x), \quad x \in [a, \eta].$$

Hence,

$$0_\star \leq \lim_{\eta \to b-0} \int_{\star a}^\eta f(x) \cdot_\star d_\star x \leq \lim_{\eta \to b-0} \int_{\star a}^\eta g(x) \cdot_\star d_\star x. \qquad (4.6)$$

Let the improper multiplicative integral (4.5) be convergent. Then

$$\lim_{\eta \to b-0} \int_{\star a}^\eta g(x) \cdot_\star d_\star x < \infty.$$

Now, applying (4.6), we get

$$0_\star \leq \lim_{\eta \to b-0} \int_{\star a}^\eta f(x) \cdot_\star d_\star x < \infty.$$

Let the improper multiplicative integral (4.4) be divergent. Then

$$\lim_{\eta \to b-0} \int_{\star a}^\eta f(x) \cdot_\star d_\star x = \infty.$$

Hence and (4.6), we conclude that

$$\lim_{\eta \to b-0} \int_{\star a}^\eta g(x) \cdot_\star d_\star x = \infty,$$

i.e., the improper multiplicative integral (4.5) is divergent. This completes the proof.

Theorem 4.6. *Let the improper multiplicative integrals* (4.4) *and* (4.5) *have a unique singularity at* $x = b$ *and*

$$0_\star < f(x), \quad 0_\star < g(x), \quad x \in [a, b].$$

Let also, there exists the limit

$$\lim_{x \to b-0} f(x)/_\star g(x) = A.$$

Then the improper multiplicative integrals (4.4) *and* (4.5) *are simultaneously convergent or divergent.*

Proof Take $\varepsilon > 0_\star$ arbitrarily. Then there exists a $c \in (a, b)$ so that

$$0_\star < (A -_\star \varepsilon) \cdot_\star g(x) < f(x) < (A +_\star \varepsilon) \cdot_\star g(x), \quad x \in [c, b).$$

Then, by the previous theorem, it follows that the improper multiplicative integrals

$$(A -_\star \varepsilon) \cdot_\star \int_{\star c}^b g(x) \cdot_\star d_\star x \quad \text{and} \quad \int_{\star c}^b g(x) \cdot_\star d_\star x$$

are simultaneously convergent or divergent. Hence, the improper multiplicative integrals

$$\int_{\star c}^b f(x) \cdot_\star d_\star x \quad \text{and} \quad \int_{\star c}^b g(x) \cdot_\star d_\star x$$

are simultaneously convergent or divergent. From here, we conclude that the improper multiplicative integrals (4.4) and (4.5) are simultaneously divergent or convergent. This completes the proof.

Example 4.8. Let the multiplicative integral (4.4) has a unique singularity at $x = b$. Let also,

$$\lim_{x \to b-0} |f(x)|_\star \cdot_\star (b -_\star x)^{\alpha_\star} = A \geq 0_\star \tag{4.7}$$

for $0_\star < \alpha < 1_\star$. Since

$$\int_{\star a}^b 1_\star /_\star (b -_\star x)^{\alpha_\star} \cdot_\star d_\star x \tag{4.8}$$

is convergent. Hence and Theorem 4.6, it follows that the multiplicative integral

$$\int_{\star a}^b |f(x)|_\star \cdot_\star d_\star x \tag{4.9}$$

is convergent. Thus, the multiplicative integral (4.4) is absolutely convergent.

Example 4.9. Let the multiplicative integral (4.4) has a unique singularity at $x = b$. Let also, (4.7) holds for $\alpha > 1_\star$ and $A > 0_\star$ and $f(x) > 0_\star$. Since (4.8) is divergent, applying Theorem 4.6, it follows that the multiplicative integral is divergent. Because $f(x) > 0_\star$, it follows that the multiplicative integral (4.4) is divergent.

Example 4.10. Let $b = \infty$ and

$$\lim_{x \to \infty} |f(x)|_\star \cdot_\star x^{\alpha_\star} = A \geq 0 \tag{4.10}$$

for $\alpha > 1_\star$. Since

$$\int_{\star a}^\infty 1_\star /_\star x^{\alpha_\star} \cdot_\star d_\star x$$

is convergent, by Theorem 4.6, it follows that the multiplicative integral (4.9) is convergent and hence, (4.4) is absolutely convergent.

Exercise 4.4. Let $b = \infty$ and (4.10) holds for $\alpha < 1$ and $A > 0$ and $f(x) > 0_\star$ for $x \in [a, \infty)$. Prove that the multiplicative integral (4.4) is divergent.

4.5   Conditional Convergence of Improper Multiplicative Integrals

Suppose that $a, b \in \mathbb{R}_\star$, $a < b$, and the multiplicative integral (4.4) has a unique singularity at $x = b$.

Theorem 4.7. *Let*

1. $f \in \mathscr{C}_\star(\mathbb{R}_\star)$.
2. $g \in \mathscr{C}_\star^1(\mathbb{R}_\star)$.
3. $F \in \mathscr{C}_\star^1([a, b))$ *be such that* $F^\star(x) = f(x)$, $x \in [a, b)$.

4. there exists

$$\int_{\ast a}^{b} F(x) \cdot_{\ast} g^{\ast}(x) \cdot_{\ast} d_{\ast}x = A.$$

5. there exists

$$\lim_{x \to b-0} F(x) \cdot_{\ast} g(x) = B.$$

Then the improper multiplicative integral

$$\int_{\ast a}^{b} f(x) \cdot_{\ast} g(x) \cdot_{\ast} d_{\ast}x$$

is convergent and

$$\int_{\ast a}^{b} f(x) \cdot_{\ast} g(x) \cdot_{\ast} d_{\ast}x = F(x) \cdot_{\ast} g(x) \Big|_{x=a}^{x=b} -_{\ast} \int_{\ast a}^{b} F(x) \cdot_{\ast} g^{\ast}(x) \cdot_{\ast} d_{\ast}x,$$

where

$$F(x) \cdot_{\ast} g(x) \Big|_{x=a}^{x=b} = \lim_{x \to b-0} F(x) \cdot_{\ast} g(x) -_{\ast} F(a) \cdot_{\ast} g(a).$$

Proof Let

$$\Phi(\eta) = \int_{\ast a}^{\eta} f(x) \cdot_{\ast} g(x) \cdot_{\ast} d_{\ast}x.$$

Then

$$\Phi(\eta) = \int_{\ast a}^{\eta} F^{\ast}(x) \cdot_{\ast} g(x) \cdot_{\ast} d_{\ast}x$$

$$= F(x) \cdot_{\ast} g(x) \Big|_{x=a}^{x=\eta} -_{\ast} \int_{\ast a}^{\eta} F(x) \cdot_{\ast} g^{\ast}(x) \cdot_{\ast} d_{\ast}x$$

and then

$$\int_{\ast a}^{b} f(x) \cdot_{\ast} g(x) \cdot_{\ast} d_{\ast}x = \lim_{\eta \to b-0} \Phi(\eta)$$

$$= \lim_{\eta \to b-0} F(x) \cdot_{\ast} g(x) \Big|_{x=a}^{x=\eta}$$

$$-_{\ast} \lim_{\eta \to b-0} \int_{\ast a}^{\eta} F(x) \cdot_{\ast} g^{\ast}(x) \cdot_{\ast} d_{\ast}x$$

$$= F(x) \cdot_{\ast} g(x) \Big|_{x=a}^{x=b}$$

$$-_{\ast} \int_{\ast a}^{b} F(x) \cdot_{\ast} g^{\ast}(x) \cdot_{\ast} d_{\ast}x.$$

This completes the proof.

Theorem 4.8. *Suppose that 1-3 of Theorem 4.7 hold. Let also,*

    1. there exists a constant $M > 0_\star$ so that

$$|F(x)|_\star \leq M, \quad x \in [a,b).$$

    2. there exists
$$\lim_{x \to b-0} g(x) = 0. \tag{4.11}$$

    3. the multiplicative integral

$$\int_{\star a}^{b} |g^\star(x)|_\star \cdot_\star d_\star x$$

is convergent.

Then the improper multiplicative integral

$$\int_{\star a}^{b} f(x) \cdot_\star g(x) \cdot_\star d_\star x$$

is convergent.

Proof We will prove that the conditions 4 and 5 of Theorem 4.7 hold. We have

$$\left| \int_{\star a}^{b} F(x) \cdot_\star g^\star(x) \cdot_\star d_\star x \right|_\star \leq \int_{\star a}^{b} |F(x)|_\star \cdot_\star |g^\star(x)|_\star \cdot_\star d_\star x$$

$$\leq M \cdot_\star \int_{\star a}^{b} |g^\star(x)|_\star \cdot_\star d_\star x.$$

Thus, the multiplicative integral

$$\int_{\star a}^{b} F(x) \cdot_\star g^\star(x) \cdot_\star d_\star x$$

is convergent, i.e., the condition 4 of Theorem 4.7 holds. Next,

$$\lim_{x \to b-0} |F(x) \cdot_\star g(x)|_\star = \lim_{x \to b-0} |F(x)|_\star \cdot_\star |g(x)|_\star$$

$$\leq M \cdot_\star \lim_{x \to b-0} |g(x)|_\star$$

$$= 0_\star,$$

i.e., the condition 5 of Theorem 4.7 holds. Now, the result follows by Theorem 4.7. This completes the proof.

4.6   The Abel-Dirichlet Criterion

Theorem 4.9. *Let*

1. $f \in \mathscr{C}_\star([a,b))$.
2. $g \in \mathscr{C}_\star^1([a,b))$.
3. $F \in \mathscr{C}_\star^1([a,b))$ *be such that* $F^\star(x) = f(x)$, $x \in [a,b)$.
4. F *be bounded on* $[a,b)$.
5. *the function g be a decreasing function on* $[a,b)$, $g \geq 0_\star$ *on* $[a,b)$,
$\lim\limits_{x \to b-0} g(x) = 0_\star$.

Then the multiplicative integral

$$\int_{\star a}^b f(x) \cdot_\star g(x) \cdot_\star d_\star x$$

is convergent.

   Proof Since g is a decreasing function on $[a,b)$, we have that $g^\star(x) \leq 0_\star$ for $x \in [a,b)$. Hence,

$$\int_{\star a}^b |g^\star(x)|_\star \cdot_\star d_\star x \quad = \quad -_\star \int_{\star a}^b g^\star(x) \cdot_\star d_\star x$$

$$= \quad g(a) -_\star \lim_{\eta \to b-0} g(\eta)$$

$$= \quad g(a).$$

Therefore, all conditions of Theorem 4.8 hold. This completes the proof.

4.7   Advanced Practical Problems

Problem 4.1. Prove that

$$\int_{\star 2}^5 e^{\frac{x}{(x-5)^{\frac{1}{2}}}} \cdot_\star d_\star x$$

is convergent.

Problem 4.2. Prove that

$$\int_{\star 4}^\infty e^{\frac{x}{(x+1)^3}} \cdot_\star d_\star x$$

is convergent.

Problem 4.3. Prove that the multiplicative integral

$$\int_{\star 2}^{10} 1_{\star}/_{\star}(10 -_{\star} x)^{15\star} \cdot_{\star} d_{\star}x$$

is divergent.

Problem 4.4. Prove that the multiplicative integral

$$\int_{\star e}^{\infty} 1_{\star}/_{\star} x^{4\star} \cdot_{\star} d_{\star}x$$

is convergent.

Problem 4.5. Prove that the multiplicative integral

$$\int_{\star 2}^{10} 1_{\star}/_{\star}((x -_{\star} 2)^{10\star} \cdot_{\star} (x -_{\star} 10)^3)$$

is convergent.

5

The Vector Space $\mathbb{R}^n_\star$

In this chapter, we introduce the space $\mathbb{R}^n_\star$ and define in it the basic multiplicative operations: multiplicative addition and multiplicative multiplication, and we prove that $\mathbb{R}^n_\star$ is a linear vector space. We define the multiplicative inner product of multiplicative vectors, multiplicative length of a multiplicative vector and multiplicative distance between two multiplicative vectors and deduct some of their properties.

5.1 Basic Definitions

Definition 5.1. We define

$$\mathbb{R}^n_\star = \{x = (x_1, x_2, \ldots, x_n) \in \mathbb{R}^n : x_j > 0, \quad j \in \{1, \ldots, n\}\}.$$

The elements of $\mathbb{R}^n_\star$ will be called multiplicative vectors.

Example 5.1. $x = (3, 4, 7, 12)$ is a multiplicative vector which is an element of $\mathbb{R}^4_\star$.

Definition 5.2. In $\mathbb{R}^n_\star$, we define the operation multiplicative addition and multiplicative multiplication of an element of $\mathbb{R}_\star$ with a multiplicative vector as follows

$$x +_\star y = (x_1 +_\star y_1, x_2 +_\star y_2, \ldots, x_n +_\star y_n)$$

and

$$\lambda \cdot_\star x = (\lambda \cdot_\star x_1, \lambda \cdot_\star x_2, \ldots, \lambda \cdot_\star x_n),$$

respectively, where

$$x = (x_1, x_2, \ldots, x_n), \quad y = (y_1, y_2, \ldots, y_n),$$

and $\lambda \in \mathbb{R}_\star$.

Example 5.2. Let

$$x = (2, 3, 4, 5),$$

$$y = (1, 3, 2, 1).$$

DOI: 10.1201/9781003299080-5

Then

$$x +_\star y = (2,3,4,5) +_\star (1,3,2,1)$$

$$= (2 +_\star 1, 3 +_\star 3, 4 +_\star 2, 5 +_\star 1)$$

$$= (2,9,8,5).$$

Example 5.3. Let

$$x = \left(\frac{1}{4}, 2, 1, \frac{1}{5}\right),$$

$$y = (4,1,2,5).$$

Then

$$x +_\star y = \left(\frac{1}{4}, 2, 1, \frac{1}{5}\right) +_\star (4,1,2,5)$$

$$= \left(\frac{1}{4} +_\star 4, 2 +_\star 1, 1 +_\star 2, \frac{1}{5} +_\star 5\right)$$

$$= (1,2,2,1).$$

Example 5.4. Let

$$x = (3,1,2),$$

$$y = (2,1,4).$$

Then

$$x +_\star y = (3,1,2) +_\star (2,1,4)$$

$$= (3 +_\star 2, 1 +_\star 1, 2 +_\star 4)$$

$$= (6,1,8).$$

Exercise 5.1. Let

$$x = \left(2,4,\frac{1}{2},3,5\right),$$

$$y = \left(3,\frac{1}{4},4,\frac{1}{3},10\right).$$

Find $x +_\star y$.

Answer 5.1.
$$(6,1,2,1,50).$$

Example 5.5. Let
$$\lambda \;=\; 3,$$

$$x \;=\; \left(\frac{1}{3},2,4,5\right).$$

Then

$$\lambda \cdot_\star x \;=\; 3 \cdot_\star \left(\frac{1}{3},2,4,5\right)$$

$$=\; \left(3 \cdot_\star \frac{1}{3}, 3 \cdot_\star 2, 3 \cdot_\star 4, 3 \cdot_5\right)$$

$$=\; \left(e^{\log 3 \log \frac{1}{3}}, e^{\log 3 \log 2}, e^{\log 3 \log 4}, e^{\log 3 \log 5}\right)$$

$$=\; \left(e^{-(\log 3)^2}, e^{\log 2 \log 3}, e^{2\log 2 \log 3} e^{\log 3 \log 5}\right).$$

Example 5.6. Let
$$\lambda \;=\; 2,$$

$$x \;=\; \left(2,\frac{1}{4},5\right).$$

Then

$$\lambda \cdot_\star x \;=\; 2 \cdot_\star \left(2,\frac{1}{4},5\right)$$

$$=\; \left(2 \cdot_\star 2, 2 \cdot_\star \frac{1}{4}, 2 \cdot_\star 5\right)$$

$$=\; \left(e^{\log 2 \log 2}, e^{\log 2 \log \frac{1}{4}}, e^{\log 2 \log 5}\right)$$

$$=\; \left(e^{(\log 2)^2}, e^{-2(\log 2)^2}, e^{\log 2 \log 5}\right).$$

Example 5.7. Let
$$\lambda \;=\; 4,$$

$$x \;=\; \left(\frac{1}{3},2,1,5,6\right).$$

Then

$$\lambda \cdot_\star x = 4 \cdot_\star \left(\frac{1}{3}, 2, 1, 5, 6 \right)$$

$$= \left(4 \cdot_\star \frac{1}{3}, 4 \cdot_\star 2, 4 \cdot_\star 1, 4 \cdot_\star 5, 4 \cdot_\star 6 \right)$$

$$= \left(e^{\log 4 \log \frac{1}{3}}, e^{\log 4 \log 2}, e^{\log 4 \log 1}, e^{\log 4 \log 5}, e^{\log 4 \log 6} \right)$$

$$= \left(e^{-2 \log 2 \log 3}, e^{2(\log 2)^2}, 1, e^{2 \log 2 \log 5}, e^{2 \log 2 (\log 2 + \log 3)} \right).$$

Exercise 5.2. Let

$$\lambda = 2,$$

$$x = \left(\frac{1}{3}, 4, 5 \right).$$

Find $\lambda \cdot_\star x$.

Answer 5.2.

$$\left(e^{-\log 2 \log 3}, e^{2(\log 2)^2}, e^{\log 2 \log 5} \right).$$

Below, we will deduct some of the properties of the defined operations. Suppose that $x = (x_1, x_2, \ldots, x_n)$, $y = (y_1, y_2, \ldots, y_n)$, $z = (z_1, z_2, \ldots, z_n) \in \mathbb{R}^n_\star$, $\lambda, \mu \in \mathbb{R}_\star$.

1. $x +_\star y = y +_\star x$.

    Proof We have

$$x +_\star y = (x_1, x_2, \ldots, x_n) +_\star (y_1, y_2, \ldots, y_n)$$

$$= (x_1 +_\star y_1, x_2 +_\star y_2, \ldots, x_n +_\star y_n)$$

$$= (y_1 +_\star x_1, y_2 +_\star x_2, \ldots, y_n +_\star x_n)$$

$$= (y_1, y_2, \ldots, y_n) +_\star (x_1, x_2, \ldots, x_n)$$

$$= y +_\star x.$$

This completes the proof.

2. $x +_\star (y +_\star z) = (x +_\star y) +_\star z$.

Proof We have

$$
\begin{aligned}
x +_\star (y +_\star z) &= (x_1, x_2, \ldots, x_n) +_\star ((y_1, y_2, \ldots, y_n) +_\star (z_1, z_2, \ldots, z_n)) \\[2mm]
&= (x_1, x_2, \ldots, x_n) +_\star (y_1 +_\star z_1, y_2 +_\star z_2, \ldots, y_n +_\star z_n) \\[2mm]
&= (x_1 +_\star (y_1 +_\star z_1), x_2 +_\star (y_2 +_\star z_2), \ldots, x_n +_\star (y_n +_\star z_n)) \\[2mm]
&= ((x_1 +_\star y_1) +_\star z_1, (x_2 +_\star y_2) +_\star z_2, \ldots, (x_n +_\star y_n) +_\star z_n) \\[2mm]
&= (x_1 +_\star y_1, x_2 +_\star y_2, \ldots, x_n +_\star y_n) +_\star (z_1, z_2, \ldots, z_n) \\[2mm]
&= (x +_\star y) +_\star z.
\end{aligned}
$$

This completes the proof.

3. $x +_\star 0_\star = x$, where $0_\star = (0, 0, \ldots, 0)$.

Proof We have

$$
\begin{aligned}
x +_\star 0_\star &= (x_1, x_2, \ldots, x_n) +_\star (0_\star, 0_\star, \ldots, 0_\star) \\[2mm]
&= (x_1 +_\star 0_\star, x_2 +_\star 0_\star, \ldots, x_n +_\star 0_\star) \\[2mm]
&= (x_1, x_2, \ldots, x_n) \\[2mm]
&= x.
\end{aligned}
$$

This completes the proof.

4. $x +_\star (-_\star x) = 0_\star$, where $-_\star x = (-_\star x_1, -_\star x_2, \ldots, -_\star x_n)$.

Proof We have

$$
\begin{aligned}
x +_\star (-_\star x) &= (x_1, x_2, \ldots, x_n) +_\star (-_\star x_1, -_\star x_2, \ldots, -_\star x_n) \\[2mm]
&= (x_1 +_\star (-_\star x_1), x_2 +_\star (-_\star x_2), \ldots, x_n +_\star (-_\star x_n)) \\[2mm]
&= (0_\star, 0_\star, \ldots, 0_\star) \\[2mm]
&= 0_\star.
\end{aligned}
$$

This completes the proof.

5. $1_\star \cdot_\star x = x.$

    Proof We have

$$1_\star \cdot_\star x \quad = \quad 1_\star \cdot_\star (x_1, x_2, \ldots, x_n)$$

$$= \quad (1_\star \cdot_\star x_1, 1_\star \cdot_\star x_2, \ldots, 1_\star \cdot_\star x_n)$$

$$= \quad (x_1, x_2, \ldots, x_n)$$

$$= \quad x.$$

This completes the proof.

6. $\lambda \cdot_\star (\mu \cdot_\star x) = (\lambda \cdot_\star \mu) \cdot_\star x.$

    Proof We have

$$\lambda \cdot_\star (\mu \cdot_\star x) \quad = \quad \lambda_\star (\mu \cdot_\star (x_1, x_2, \ldots, x_n))$$

$$= \quad \lambda \cdot_\star (\mu \cdot_\star x_1, \mu \cdot_\star x_2, \ldots, \mu \cdot_\star x_n)$$

$$= \quad (\lambda \cdot_\star (\mu \cdot_\star x_1), \lambda \cdot_\star (\mu \cdot_\star x_2), \ldots, \lambda \cdot_\star (\mu \cdot_\star x_n))$$

$$= \quad ((\lambda \cdot_\star \mu) \cdot_\star x_1, (\lambda \cdot_\star \mu) \cdot_\star x_2, \ldots, (\lambda \cdot_\star \mu) \cdot_\star x_n)$$

$$= \quad (\lambda \cdot_\star \mu) \cdot_\star (x_1, x_2, \ldots, x_n)$$

$$= \quad (\lambda \cdot_\star \mu) \cdot_\star x.$$

This completes the proof.

7. $(\lambda +_\star \mu) \cdot_\star x = \lambda \cdot_\star x +_\star \mu_\star \cdot_\star x.$

    Proof We have

$$(\lambda +_\star \mu) \cdot_\star x \quad = \quad (\lambda +_\star \mu) \cdot_\star (x_1, x_2, \ldots, x_n)$$

$$= \quad ((\lambda +_\star \mu) \cdot_\star x_1, (\lambda +_\star \mu) \cdot_\star x_2, \ldots, (\lambda +_\star \mu) \cdot_\star x_n)$$

$$= \quad (\lambda \cdot_\star x_1 +_\star \mu \cdot_\star x_1, \lambda \cdot_\star x_2 +_\star \mu \cdot_\star x_2, \ldots, \lambda \cdot_\star x_n +_\star \mu \cdot_\star x_n)$$

$$= \quad (\lambda \cdot_\star x_1, \lambda \cdot_\star x_2, \ldots, \lambda \cdot_\star x_n)$$

$$+_\star (\mu \cdot_\star x_1, \mu \cdot_\star x_2, \ldots, \mu \cdot_\star x_n)$$

$$= \quad \lambda \cdot_\star (x_1, x_2, \ldots, x_n)$$

$$+_\star \mu \cdot_\star (x_1, x_2, \ldots, x_n)$$

$$= \quad \lambda \cdot_\star x +_\star \mu \cdot_\star x.$$

This completes the proof.

8. $\lambda \cdot_\star (x +_\star y) = \lambda \cdot_\star x +_\star \lambda \cdot_\star y.$

Proof We have

$$\lambda \cdot_\star (x +_\star y) \quad = \quad \lambda \cdot_\star ((x_1, x_2, \ldots, x_n) +_\star (y_1, y_2, \ldots, y_n))$$

$$= \quad \lambda \cdot_\star (x_1 +_\star y_1, x_2 +_\star y_2, \ldots, x_n +_\star y_n)$$

$$= \quad (\lambda \cdot_\star (x_1 +_\star y_1), \lambda \cdot_\star (x_2 +_\star y_2), \ldots, \lambda \cdot_\star x_n +_\star y_n)$$

$$= \quad (\lambda \cdot_\star x_1 +_\star \lambda \cdot_\star y_1, \lambda \cdot_\star x_2 +_\star \lambda \cdot_\star y_2, \ldots, \lambda \cdot_\star x_n +_\star \lambda \cdot_\star y_n)$$

$$= \quad (\lambda \cdot_\star x_1, \lambda \cdot_\star x_2, \ldots, \lambda \cdot_\star x_n)$$

$$+_\star (\lambda \cdot_\star y_1, \lambda \cdot_\star y_2, \ldots, \lambda \cdot_\star y_n)$$

$$= \quad \lambda \cdot_\star (x_1, x_2, \ldots, x_n) +_\star \lambda \cdot_\star (y_1, y_2, \ldots, y_n)$$

$$= \quad \lambda \cdot_\star x +_\star \lambda \cdot_\star y.$$

This completes the proof.

Therefore, $\mathbb{R}^n_\star$ is a linear vector space.

Example 5.8. Let

$$x \quad = \quad (3, 1, 4),$$

$$y \quad = \quad (2, 3, 1).$$

We will find

$$z = (3 \cdot_\star x -_\star y) +_\star 2 \cdot_\star x.$$

We have

$$
\begin{aligned}
3 \cdot_\star x &= 3 \cdot_\star (3,1,4) \\
&= (3 \cdot_\star 3, 3 \cdot_\star 1, 3 \cdot_\star 4) \\
&= \left(e^{(\log 3)^2}, e^{\log 3 \log 1}, e^{\log 3 \log 4} \right) \\
&= \left(e^{(\log 3)^2}, 1, e^{2 \log 3 \log 2} \right)
\end{aligned}
$$

and

$$
\begin{aligned}
3 \cdot_\star x -_\star y &= \left(e^{(\log 3)^2}, 1, e^{2 \log 3 \log 2} \right) \\
&\quad -_\star (2,3,1) \\
&= \left(e^{(\log 3)^2} -_\star 2, 1 -_\star 3, e^{2 \log 3 \log 2} -_\star 1 \right) \\
&= \left(\frac{1}{2} e^{(\log 3)^2}, \frac{1}{3}, e^{2 \log 3 \log 2} \right),
\end{aligned}
$$

and

$$
\begin{aligned}
2 \cdot_\star x &= 2 \cdot_\star (3,1,4) \\
&= (2 \cdot_\star 3, 2 \cdot_\star 1, 2 \cdot_\star 4) \\
&= \left(e^{\log 2 \log 3}, e^{\log 2 \log 1}, e^{\log 2 \log 4} \right) \\
&= \left(e^{\log 2 \log 3}, 1, e^{2(\log 2)^2} \right).
\end{aligned}
$$

Hence,

$$
\begin{aligned}
z &= \left(\frac{1}{2} e^{(\log 3)^2}, \frac{1}{3}, e^{2 \log 3 \log 2} \right) \\
&\quad +_\star \left(e^{\log 2 \log 3}, 1, e^{2(\log 2)^2} \right) \\
&= \left(\frac{1}{2} e^{(\log 3)^2} +_\star e^{\log 2 \log 3}, \frac{1}{3} +_\star 1, e^{2 \log 3 \log 2} +_\star e^{2(\log 2)^2} \right)
\end{aligned}
$$

$$= \left(\frac{1}{2} e^{\log 3 (\log 2 + \log 3)}, \frac{1}{3} e^{2 \log 2 (\log 2 + \log 3)} \right).$$

Exercise 5.3. Let

$$x \;=\; (2,3),$$

$$y \;=\; (4,2),$$

$$z \;=\; (2,1).$$

Find

$$(2 \cdot_\star x +_\star y) -_\star z.$$

Answer 5.3.

$$\left(2 e^{(\log 2)^2}, 2 e^{\log 2 \log 3} \right).$$

5.2 Multiplicative Linear Dependence and Independence

Definition 5.3. We say that the system of multiplicative vectors $e_1, e_2, \ldots, e_r \in \mathbb{R}_\star^n$ is multiplicative linearly dependent if there are $\lambda_1, \lambda_2, \ldots, \lambda_r \in \mathbb{R}_\star$ so that

$$(\lambda_1, \lambda_2, \ldots, \lambda_r) \neq (0_\star, 0_\star, \ldots, 0_\star)$$

and

$$\lambda_1 \cdot_\star e_1 +_\star \lambda_2 \cdot_\star e_2 +_\star \cdots +_\star \lambda_r \cdot_\star e_r = 0_\star.$$

Otherwise, it is said to be multiplicative linearly independent.

Let

$$e_1 \;=\; (e_{11}, e_{21}, \ldots, e_{n1}),$$

$$e_2 \;=\; (e_{12}, e_{22}, \ldots, e_{n2}),$$

$$\vdots$$

$$e_r \;=\; (e_{1r}, e_{2r}, \ldots, e_{nr}).$$

We have

$$\lambda_1 \cdot_\star e_1 +_\star \lambda_2 \cdot_\star e_2 +_\star \cdots +_\star \lambda_r \cdot_\star e_r$$

$$= \quad \lambda_1 \cdot_\star (e_{11}, e_{21}, \ldots, e_{n1}) +_\star \lambda_2 \cdot_\star (e_{12}, e_{22}, \ldots, e_{n2})$$

$$+_\star \cdots$$

$$+_\star \lambda_r \cdot_\star (e_{1r}, e_{2r}, \ldots, e_{nr})$$

$$= \quad (\lambda_1 \cdot_\star e_{11}, \lambda_1 \cdot_\star e_{21}, \ldots, \lambda_1 \cdot_\star e_{n1})$$

$$+_\star (\lambda_2 \cdot_\star e_{12}, \lambda_2 \cdot_\star e_{22}, \ldots, \lambda_2 \cdot_\star e_{n2})$$

$$+_\star \cdots$$

$$+_\star (\lambda_r \cdot_\star e_{1r}, \lambda_r \cdot_\star e_{2r}, \ldots, \lambda_r \cdot_\star e_{nr})$$

$$= \quad (\lambda_1 \cdot_\star e_{11} +_\star \lambda_2 \cdot_\star e_{12} +_\star \cdots +_\star \lambda_r \cdot_\star e_{1r},$$

$$\lambda_1 \cdot_\star e_{21} +_\star \lambda_2 \cdot_\star e_{22} +_\star \cdots +_\star \lambda_r \cdot_\star e_{2r},$$

$$\ldots,$$

$$\lambda_1 \cdot_\star e_{n1} +_\star \lambda_2 \cdot_\star e_{n2} +_\star \cdots +_\star \lambda_r \cdot_\star e_{nr})$$

$$= \quad \left(e^{\log \lambda_1 \log e_{11} + \log \lambda_2 \log e_{12} + \cdots + \log \lambda_r \log e_{1r}}, \right.$$

$$e^{\log \lambda_1 \log e_{21} + \log \lambda_2 \log e_{22} + \cdots + \log \lambda_r \log e_{2r}},$$

$$\ldots,$$

$$\left. e^{\log \lambda_1 \log e_{n1} + \log \lambda_2 \log e_{n2} + \cdots + \log \lambda_r \log e_{nr}} \right).$$

Then

$$\lambda_1 \cdot_\star e_1 +_\star \lambda_2 \cdot_\star e_2 +_\star \cdots +_\star \lambda_r \cdot_\star e_r = 0_\star$$

if and only if

$$0 = \log \lambda_1 \log e_{11} + \log \lambda_2 \log e_{12} + \cdots + \log \lambda_r \log e_{1r}$$

$$0 = \log \lambda_1 \log e_{21} + \log \lambda_2 \log e_{22} + \cdots + \log \lambda_r \log e_{2r}$$

$$\vdots$$

$$0 \;=\; \log \lambda_1 \log e_{n1} + \log \lambda_2 \log e_{n2} + \cdots + \log \lambda_r \log e_{nr}.$$

Example 5.9. Let

$$e_1 \;=\; (2,4)$$

$$e_2 \;=\; (4,8).$$

Then

$$0 \;=\; \log \lambda_1 \log 2 + \log \lambda_2 \log 4$$

$$0 \;=\; \log \lambda_1 \log 4 + \log \lambda_2 \log 8$$

or

$$0 \;=\; (\log \lambda_1 + 2\log \lambda_2) \log 2$$

$$0 \;=\; (2\log \lambda_1 + 3\log \lambda_2) \log 2,$$

or

$$0 \;=\; \log \lambda_1 + 2\log \lambda_2$$

$$0 \;=\; 2\log \lambda_1 + 3\log \lambda_2.$$

Thus,

$$\log \lambda_1 \;=\; 0$$

$$\log \lambda_2 \;=\; 0$$

or

$$\lambda_1 \;=\; 1$$

$$\lambda_2 \;=\; 1.$$

Therefore, the multiplicative vectors e_1 and e_2 are multiplicative linearly independent and linearly dependent.

Example 5.10. Let

$$e_1 = \left(e^2, e^4 \right)$$

$$e_2 = \left(e^4, e^8 \right).$$

Take

$$\lambda_1 = e^2, \quad \lambda_2 = -_\star e.$$

Then

$$\log \lambda_1 \log e^2 + \log \lambda_2 \log e^4 \;=\; 4 - 4$$

$$=\; 0,$$

$$\log \lambda_1 \log e^4 + \log \lambda_2 \log e^8 \;=\; 8 - 8$$

$$=\; 0.$$

Also,

$$(\lambda_1, \lambda_2) \;=\; \left(e^2, \frac{1}{2} \right)$$

$$\neq\; (0_\star, 0_\star).$$

Therefore, e_1 and e_2 are multiplicative linearly dependent and linearly independent.

Example 5.11. Let

$$e_1 = (2,3)$$

$$e_2 = (1,4).$$

Then

$$\log \lambda_1 \log 2 + \log \lambda_2 \log 1 \;=\; 0$$

$$\log \lambda_1 \log 3 + \log \lambda_2 \log 4 \;=\; 0,$$

whereupon

$$\log \lambda_1 \;=\; 0$$

$$\log \lambda_2 \;=\; 0$$

or

$$\lambda_1 \quad = \quad 0_\star$$

$$\lambda_2 \quad = \quad 0_\star.$$

Thus, the considered multiplicative vectors are multiplicative linearly independent and linearly independent.

Exercise 5.4. Prove that

$$e_1 \quad = \quad \left(e, e^2, e^3\right)$$

$$e_2 \quad = \quad \left(e^2, e^4, e^6\right)$$

$$e_3 \quad = \quad \left(e^4, e^8, e^{12}\right)$$

are multiplicative linearly dependent and linearly independent.

Exercise 5.5. Prove that

$$e_1 \quad = \quad (2, 4, 6)$$

$$e_2 \quad = \quad (4, 8, 12)$$

$$e_3 \quad = \quad (8, 16, 24)$$

are multiplicative linearly independent and linearly dependent.

5.3 Multiplicative Inner Product

Let

$$x = (x_1, x_2, \ldots, x_n), \quad y = (y_1, y_2, \ldots, y_n) \in \mathbb{R}_\star^n.$$

Definition 5.4. We define the multiplicative inner product of x and y as follows

$$\langle x, y \rangle_\star \quad = \quad x_1 \cdot_\star y_1 +_\star x_2 \cdot_\star y_2 +_\star \cdots +_\star x_n \cdot_\star y_n$$

$$= \quad e^{\log x_1 \log y_1 + \log x_2 \log y_2 + \cdots + \log x_n \log y_n}.$$

Example 5.12. Let

$$x = (2,1,3),$$

$$y = (e,2,1).$$

Then

$$\langle x,y\rangle_\star = e^{\log 2\log e + \log 1\log 2 + \log 3\log 1}$$

$$= e^{\log 2}$$

$$= 2.$$

Example 5.13. Let

$$x = (3,2,e,4),$$

$$y = (e,3,4,e).$$

Then

$$\langle x,y\rangle_\star = e^{\log 3\log e + \log 2\log 3 + \log e\log 4 + \log 4\log e}$$

$$= e^{\log 3 + \log 2\log 3 + 4\log 2}.$$

Example 5.14. Let

$$x = (3,4,e),$$

$$y = (2,2,3).$$

Then

$$\langle x,y\rangle_\star = e^{\log 3\log 2 + \log 4\log 2 + \log e\log 3}$$

$$= e^{\log 3\log 2 + 2(\log 2)^2 + \log 3}.$$

Exercise 5.6. Let

$$x = (e^2,e,e^3),$$

$$y = (e,e^2,e^3).$$

Find $\langle x,y\rangle_\star$.

Answer 5.4. e^{13}.

Theorem 5.1. *Let*

$$x = (x_1, x_2, \ldots, x_n), \quad y = (y_1, y_2, \ldots, y_n), \quad z = (z_1, z_2, \ldots, z_n) \in \mathbb{R}_\star^n.$$

Then

$$\langle x, y +_\star z \rangle_\star = \langle x, y \rangle_\star +_\star \langle x, z \rangle_\star.$$

Proof We have

$$
\begin{aligned}
y +_\star z &= (y_1, y_2, \ldots, y_n) +_\star (z_1, z_2, \ldots, z_n) \\[1em]
&= (y_1 +_\star z_1, y_2 +_\star z_2, \ldots, y_n +_\star z_n) \\[1em]
&= (y_1 z_1, y_2 z_2, \ldots, y_n z_n), \\[1em]
\langle x, y \rangle_\star &= e^{\log x_1 \log y_1 + \log x_2 \log y_2 + \cdots + \log x_n \log y_n}, \\[1em]
\langle x, z \rangle_\star &= e^{\log x_1 \log z_1 + \log x_2 \log z_2 + \log x_n \log z_n}.
\end{aligned}
$$

Hence,

$$\langle x, y \rangle_\star +_\star \langle x, z \rangle_\star = e^{\log x_1 \log y_1 + \log x_2 \log y_2 + \cdots + \log x_n \log y_n} e^{\log x_1 \log z_1 + \log x_2 \log z_2 + \cdots + \log x_n \log z_n}$$

$$= e^{\log x_1 \log y_1 + \log x_2 \log y_2 + \cdots + \log x_n \log y_n + \log x_1 \log z_1 + \log x_2 \log z_2 + \cdots + \log x_n \log z_n}$$

$$= e^{\log x_1 (\log y_1 + \log z_1) + \log x_2 (\log y_2 + \log z_2) + \cdots + \log x_n (\log y_n + \log z_n)}$$

$$= e^{\log x_1 \log(y_1 z_1) + \log x_2 \log(y_2 z_2) + \log x_n (\log y_n \log z_n)}$$

$$= \langle x, y +_\star z \rangle_\star.$$

This completes the proof.

Theorem 5.2. *Let*

$$x = (x_1, x_2, \ldots, x_n), \quad y = (y_1, y_2, \ldots, y_n) \in \mathbb{R}_\star^n, \quad c \in \mathbb{R}_\star.$$

Then

$$c \cdot_\star \langle x, y \rangle_\star = \langle c \cdot_\star x, y \rangle_\star = \langle x, c \cdot_\star y \rangle_\star. \tag{5.1}$$

Proof We have

$$c \cdot_\star x = c \cdot_\star (x_1, x_2, \ldots, x_n)$$

$$= (c \cdot_\star x_1, c \cdot_\star x_2, \ldots, c \cdot_\star x_n)$$

$$= \left(e^{\log c \log x_1}, e^{\log c \log x_2}, \ldots, e^{\log c \log x_n} \right)$$

and as above,

$$c \cdot_\star y = \left(e^{\log c \log y_1}, e^{\log c \log y_2}, \ldots, e^{\log c \log y_n} \right).$$

Next,

$$\langle x, y \rangle_\star = e^{\log x_1 \log y_1 + \log x_2 \log y_2 + \cdots + \log x_n \log y_n}$$

and

$$
\begin{aligned}
c \cdot_\star \langle x, y \rangle_\star &= c \cdot_\star e^{\log x_1 \log y_1 + \log x_2 \log y_2 + \cdots + \log x_n \log y_n} \\
&= e^{\log c e^{\log x_1 \log y_1 + \log x_2 \log y_2 + \cdots + \log x_n \log y_n}} \qquad (5.2) \\
&= e^{\log c (\log x_1 \log y_1 + \log x_2 \log y_2 + \cdots + \log x_n \log y_n)}
\end{aligned}
$$

and

$$
\begin{aligned}
\langle c \cdot_\star x, y \rangle_\star &= e^{\left(\log e^{\log c \log x_1} \right) \log y_1 + \left(\log e^{\log c \log x_2} \right) \log y_2 + \cdots + \left(\log e^{\log c \log x_n} \right) \log y_n} \\
&= e^{\log c \log x_1 \log y_1 + \log c \log x_2 \log y_2 + \cdots + \log c \log x_n \log y_n} \qquad (5.3) \\
&= e^{\log c (\log x_1 \log y_1 + \log x_2 \log y_2 + \cdots + \log x_n \log y_n)},
\end{aligned}
$$

and

$$
\begin{aligned}
\langle x, c \cdot_\star y \rangle_\star &= e^{\log x_1 \log e^{\log c \log y_1} + \log x_2 \log e^{\log c \log y_2} + \cdots + \log x_n \log e^{\log c \log y_n}} \\
&= e^{\log x_1 \log c \log y_1 + \log x_2 \log c \log y_2 + \cdots + \log x_n \log c \log y_n} \qquad (5.4) \\
&= e^{\log c (\log x_1 \log y_1 + \log x_2 \log y_2 + \cdots + \log x_n \log y_n)}.
\end{aligned}
$$

By equations (5.2), (5.3) and (5.4), we get equation (5.1). This completes the proof.

Corollary 5.1. Suppose that all conditions of Theorem 5.2 hold. Then

$$\langle x, y \rangle_\star = \langle y, x \rangle_\star.$$

Proof We have

$$
\begin{aligned}
\langle y, x \rangle_\star &= e^{\log y_1 \log x_1 + \log y_2 \log x_2 + \star \cdots +_\star \log y_n \log x_n} \\
&= e^{\log x_1 \log y_1 + \log x_2 \log y_2 + \star \cdots +_\star \log x_n \log y_n} \\
&= \langle x, y \rangle_\star.
\end{aligned}
$$

This completes the proof.

Theorem 5.3. *If* $\langle x,y \rangle_\star = 0_\star$ *for any* $x \in \mathbb{R}^n_\star$, *then* $y = 0_\star$.

Proof We have

$$0_\star \;=\; 1$$

$$=\; \langle x,y \rangle_\star$$

$$=\; e^{\log x_1 \log y_1 + \log x_2 \log y_2 + \cdots + \log x_n \log y_n}$$

$$=\; e^0$$

for any $x_1,x_2,\ldots,x_n \in \mathbb{R}_\star$. For $x_1 = x_2 = \ldots = x_n = e$, we get

$$e^0 \;=\; e^{\log y_1 + \log y_2 + \cdots + \log y_n}$$

$$=\; e^{\log(y_1 y_2 \cdots y_n)},$$

whereupon

$$y_1 y_2 \ldots y_n = 1. \tag{5.5}$$

For $x_1 = e, x_j = e^{-1}, x_l = 0_\star, j,l \in \{2,\ldots,n\}, j \neq l$. we find

$$e^{\log y_1 - \log y_j} = e^0$$

or

$$e^{\log \frac{y_1}{y_j}} = e^0.$$

Therefore,

$$\frac{y_1}{y_j} = 1, \quad j \in \{1,2,\ldots,n\}.$$

By the last equation and equation (5.5), we arrive at the system

$$\frac{y_1}{y_j} \;=\; 1, \quad j \in \{1,\ldots,n\},$$

$$y_1 y_2 \ldots y_n \;=\; 1.$$

Hence,

$$y_1 = y_2 = \cdots = y_n = 1$$

and

$$y \;=\; (y_1,y_2,\ldots,y_n)$$

$$=\; (1,1,\ldots,1)$$

$$=\; 0_\star.$$

This completes the proof.

Definition 5.5. Two multiplicative vectors $x, y \in \mathbb{R}_\star^n$ are said to be multiplicative orthogonal if

$$\langle x, y \rangle_\star = 0_\star.$$

We will write $x \perp_\star y$.

Example 5.15. Let

$$x = (e, e^2, e),$$

$$y = (e^{-1}, e, e^{-1}).$$

Then

$$
\begin{aligned}
\langle x, y \rangle_\star &= e^{\log e \log e^{-1} + \log e^2 \log e + \log e \log e^{-1}} \\
&= e^{-1 + 2 - 1} \\
&= e^0 \\
&= 1 \\
&= 0_\star.
\end{aligned}
$$

Thus, $x \perp_\star y$. On the other hand,

$$
\begin{aligned}
\langle x, y \rangle &= 1 + e^3 + 1 \\
&= 2 + e^3 \\
&\neq 0.
\end{aligned}
$$

Therefore, $x \not\perp y$.

Exercise 5.7. Let

$$x = (e, e^3, e^4, e^5),$$

$$y = (e^2, e, e^2, e^3).$$

Check if $x \perp_\star y$.

Answer 5.5. *No.*

5.4   Multiplicative Length and Multiplicative Distance

Definition 5.6. For $x \in \mathbb{R}_\star^n$, $x = (x_1, x_2, \ldots, x_n)$, we define the multiplicative length as follows

$$|x|_\star = e^{((\log x_1)^2 + (\log x_2)^2 + \cdots + (\log x_n)^2)^{\frac{1}{2}}}.$$

Example 5.16. Let

$$x = (e, e^2, e^3, e^4, e^{-1}).$$

Then

$$
\begin{aligned}
|x|_\star &= e^{((\log e)^2 + (\log e^2)^2 + (\log e^3)^2 + (\log e^4)^2 + (\log e^{-1})^2)^{\frac{1}{2}}} \\[2mm]
&= e^{(1+4+9+16+1)^{\frac{1}{2}}} \\[2mm]
&= e^{\sqrt{31}}.
\end{aligned}
$$

Example 5.17. Let

$$x = (e^{-2}, e, e).$$

Then

$$
\begin{aligned}
|x|_\star &= e^{((\log e^{-2})^2 + (\log e)^2 + (\log e)^2)^{\frac{1}{2}}} \\[2mm]
&= e^{(4+1+1)^{\frac{1}{2}}} \\[2mm]
&= e^{\sqrt{6}}.
\end{aligned}
$$

Example 5.18. Let

$$x = (2, 4, 8).$$

Then

$$
\begin{aligned}
|x|_\star &= e^{((\log 2)^2 + (\log 4)^2 + (\log 8)^2)^{\frac{1}{2}}} \\[2mm]
&= e^{((\log 2)^2 + 4(\log 2)^2 + 9(\log 2)^2)^{\frac{1}{2}}} \\[2mm]
&= e^{\sqrt{14}\log 2}.
\end{aligned}
$$

Exercise 5.8. Let

$$x = (e, e^{-1}, e^{-1}, e^{-1}).$$

Find $|x|_\star$.

Answer 5.6. e^2.

Definition 5.7. For $x, y \in \mathbb{R}_\star^n$, we define the multiplicative distance between x and y as follows

$$|x -_\star y|_\star.$$

For $x = (x_1, x_2, \ldots, x_n)$, $y = (y_1, y_2, \ldots, y_n)$, we define

$$
\begin{aligned}
|x -_\star y|_\star &= |(x_1, x_2, \ldots, x_n) -_\star (y_1, y_2, \ldots, y_n)|_\star \\[2mm]
&= (x_1 -_\star y_1, x_2 -_\star y_2, \ldots, x_n -_\star y_n)|_\star \\[2mm]
&= \left| \left(\frac{x_1}{y_1}, \frac{x_2}{y_2}, \ldots, \frac{x_n}{y_n} \right) \right|_\star \\[2mm]
&= e^{\left((\log x_1 - \log y_1)^2 + (\log x_2 - \log y_2)^2 + \cdots + (\log x_n - \log y_n)^2 \right)^{\frac{1}{2}}}.
\end{aligned}
$$

Example 5.19. Let

$$x = (2, e^2, e^3),$$

$$y = (2, e, e).$$

Then

$$
\begin{aligned}
x -_\star y &= (2, e^2, e^3) -_\star (2, e, e) \\[2mm]
&= (2 -_\star 2, e^2 -_\star e, e^3 -_\star e) \\[2mm]
&= \left(\frac{2}{2}, \frac{e^2}{e}, \frac{e^3}{e} \right) \\[2mm]
&= (1, e, e^2)
\end{aligned}
$$

and hence,

$$
\begin{aligned}
|x -_\star y|_\star &= e^{\left((\log 1)^2 + (\log e)^2 + (\log e^2)^2 \right)^{\frac{1}{2}}} \\[2mm]
&= e^{(1+4)^{\frac{1}{2}}} \\[2mm]
&= e^{\sqrt{5}}.
\end{aligned}
$$

Example 5.20. Let

$$x = (2,4,8),$$

$$y = (4,8,16).$$

Then

$$x -_\star y = (2,4,8) -_\star (4,8,16)$$

$$= (2 -_\star 4, 4 -_\star 8, 8 -_\star 16)$$

$$= \left(\frac{2}{4}, \frac{4}{8}, \frac{8}{16}\right)$$

$$= \left(\frac{1}{2}, \frac{1}{2}, \frac{1}{2}\right).$$

Hence,

$$|x -_\star y|_\star = e^{\left((\log 2)^2 + (\log 2)^2 + (\log 2)^2\right)^{\frac{1}{2}}}$$

$$= e^{\sqrt{3}\log 2}.$$

Example 5.21. Let

$$x = (2,4,8),$$

$$y = (1,1,1).$$

Then

$$x -_\star y = (2,4,8) -_\star (1,1,1)$$

$$= (2 -_\star 1, 4 -_\star 1, 8 -_\star 1)$$

$$= (2,4,8)$$

and

$$|x -_\star y|_\star = e^{\left((\log 2)^2 + (\log 4)^2 + (\log 8)^2\right)^{\frac{1}{2}}}$$

$$= e^{\left((\log 2)^2 + 4(\log 2)^2 + 9(\log 2)^2\right)^{\frac{1}{2}}}$$

$$= e^{\sqrt{14}\log 2}.$$

Exercise 5.9. Let

$$x = (3, 1, 2),$$

$$y = (4, 2, 1).$$

Find $|x -_\star y|_\star$.

Answer 5.7.

$$e^{\left(\left(\log \frac{3}{4}\right)^2 + 2(\log 2)^2\right)^{\frac{1}{2}}}.$$

Note that

$$\langle x, x \rangle_\star = e^{(\log x_1)^2 + (\log x_2)^2 + \cdots + (\log x_n)^2}$$

and

$$(\langle x, x \rangle_\star)^{\frac{1}{2}\star} = e^{\left(\log e^{(\log x_1)^2 + (\log x_2)^2 + \cdots + (\log x_n)^2}\right)^{\frac{1}{2}}}$$

$$= |x|_\star.$$

Theorem 5.4. *The multiplicative length has the following properties.*

    1. $|x|_\star \geq_\star 0_\star$.
    2. If $|x|_\star = 0_\star$*, then* $x = 0_\star$.
    3. $|c \cdot_\star x|_\star = |c|_\star \cdot_\star |x|_\star$.

Proof

  1. We have

$$|x|_\star = e^{\left((\log x_1)^2 + (\log x_2)^2 + \cdots + (\log x_n)^2\right)^{\frac{1}{2}}}$$

$$\geq e^0$$

$$= 1$$

$$= 0_\star.$$

  2. Let $|x|_\star = 0_\star$. Then

$$e^{\left((\log x_1)^2 + (\log x_2)^2 + \cdots + (\log x_n)^2\right)^{\frac{1}{2}}} = 0_\star$$

$$= 1$$

$$= e^0.$$

Hence,

$$(\log x_1)^2 + (\log x_2)^2 + \cdots + (\log x_n)^2 = 0$$

and

$$\log x_1 = 0$$

$$\log x_2 = 0$$

$$\vdots$$

$$\log x_n = 0.$$

Therefore,

$$x_1 = x_2 = \ldots = x_n = 1$$

and $x = 0_\star$.

3. Let $c \geq 1$. Then $|c|_\star = c$ and

$$c \cdot_\star x = \left(e^{\log c \log x_1}, e^{\log c \log x_2}, \ldots, e^{\log c \log x_n} \right),$$

and

$$|c \cdot_\star x|_\star = e^{\left(\left(\log e^{\log e^{\log c \log x_1}} \right)^2 + \left(\log e^{\log e^{\log c \log x_2}} \right)^2 \right)^{\frac{1}{2}} + (\log c \log x_n)^2}$$

$$= e^{\left((\log c \log x_1)^2 + (\log c \log x_2)^2 + \cdots + (\log c \log x_n)^2 \right)^{\frac{1}{2}}}$$

$$= e^{\log c \left((\log x_1)^2 + (\log x_2)^2 + \cdots + (\log x_n)^2 \right)^{\frac{1}{2}}},$$

and

$$c \cdot_\star |x|_\star = c \cdot_\star e^{\left((\log x_1)^2 + (\log x_2)^2 + \cdots + (\log x_n)^2 \right)^{\frac{1}{2}}}$$

$$= e^{\log c \log e^{\left((\log x_1)^2 + (\log x_2)^2 + \cdots + (\log x_n)^2 \right)^{\frac{1}{2}}}}$$

$$= e^{\log c \left((\log x_1)^2 + (\log x_2)^2 + \cdots + (\log x_n)^2 \right)^{\frac{1}{2}}}.$$

Therefore,

$$|c \cdot_\star x|_\star = c \cdot_\star |x|_\star.$$

The case $c \in (0,1]$ we leave to the reader as an exercise. This completes the proof.

Theorem 5.5 (The Multiplicative Cauchy-Schwartz Inequality). *For any $x,y \in \mathbb{R}_\star^n$, we have*

$$|\langle x,y \rangle_\star|_\star \leq |x|_\star \cdot_\star |y|_\star.$$

The equality holds if and only if x and y are multiplicative proportional.

Proof Let $x = (x_1, x_2, \ldots, x_n)$ and $y = (y_1, y_2, \ldots, y_n)$. Then

$$\langle x,y \rangle_\star = e^{\log x_1 \log y_1 + \log x_2 \log y_2 + \cdots + \log x_n \log y_n}$$

and

$$|\langle x,y \rangle_\star|_\star = \begin{cases} e^{\log x_1 \log y_1 + \log x_2 \log y_2 + \cdots + \log x_n \log y_n} \\ \quad \text{if} \quad \log x_1 \log y_1 + \log x_2 \log y_2 + \cdots + \log x_n \log y_n \geq 0 \\ e^{-(\log x_1 \log y_1 + \log x_2 \log y_2 + \cdots + \log x_n \log y_n)} \\ \quad \text{if} \quad \log x_1 \log y_1 + \log x_2 \log y_2 + \cdots + \log x_n \log y_n < 0. \end{cases}$$

$$\leq e^{\left((\log x_1)^2 + (\log x_2)^2 + \cdots + (\log x_n)^2\right)^{\frac{1}{2}} \left((\log y_1)^2 + (\log y_2)^2 + \cdots + (\log y_n)^2\right)^{\frac{1}{2}}}$$

$$= |x|_\star \cdot_\star |y|_\star.$$

Next,

$$|\langle x,y \rangle|_\star = |x|_\star \cdot_\star |y|_\star$$

if and only if

$$e^{|\log x_1 \log y_1 + \log x_2 \log y_2 + \cdots + \log x_n \log y_n|}$$
$$= e^{\left((\log x_1)^2 + (\log x_2)^2 + \cdots + (\log x_n)^2\right)^{\frac{1}{2}} \left((\log y_1)^2 + (\log y_2)^2 + \cdots + (\log y_n)^2\right)^{\frac{1}{2}}},$$

if and only if

$$(\log x_1 \log y_1 + \log x_2 \log y_2 + \cdots + \log x_n \log y_n)^2$$
$$= \left((\log x_1)^2 + (\log x_2)^2 + \cdots + (\log x_n)^2\right)$$
$$\times \left((\log y_1)^2 + (\log y_2)^2 + \cdots + (\log y_n)^2\right)$$

if and only if

$$(\log x_1)^2 (\log y_1)^2 + (\log x_2)^2 (\log y_2)^2 + \cdots + (\log x_n)^2 (\log y_n)^2$$

$$+2\sum_{i<j} \log x_i \log x_j \log y_i \log y_j$$

$$= \sum_{i=1}^{n}(\log x_i)^2(\log y_i)^2 + \sum_{i,j=1,i\neq j}^{n}(\log x_i)^2(\log y_j)^2,$$

if and only if

$$e^{\log x_i \log y_j} = e^{\log x_j \log y_i}, \quad i,j \in \{1,\dots,n\},$$

if and only if

$$x_i \cdot_\star y_j = x_j \cdot_\star y_i, \quad i,j \in \{1,\dots,n\},$$

if and only if

$$x_i/_\star y_i = x_j/_\star y_j, \quad i,j \in \{1,\dots,n\}.$$

This completes the proof.

Corollary 5.2. For any $x,y \in E_\star^n$, we have

$$|x+_\star y|_\star \leq |x|_\star \cdot_\star |y|_\star.$$

Proof We have, using the multiplicative Cauchy-Schwartz inequality,

$$|x+_\star y|_\star^{2\star} = \langle x+_\star y, x+_\star y\rangle_\star$$

$$= \langle x,x\rangle_\star +_\star e^2 \cdot_\star \langle x,y\rangle_\star +_\star \langle y,y\rangle_\star$$

$$= |x|_\star^{2\star} +_\star e^2 \cdot_\star \langle x,y\rangle_\star +_\star |y|_\star^{2\star}$$

$$\leq |x|_\star^{2\star} +_\star e^2 \cdot_\star |x|_\star \cdot_\star |y|_\star +_\star |y|_\star^{2\star}$$

$$= (|x|_\star +_\star |y|_\star)^{2\star},$$

whereupon

$$|x+_\star y|_\star \leq |x|_\star +_\star |y|_\star.$$

This completes the proof.

5.5   Advanced Practical Problems

Problem 5.1. Let

$$x = \left(3,2,\frac{1}{3},4,5\right),$$

$$y = \left(\frac{1}{3}, 4, 6, 1, \frac{1}{10}\right).$$

Find $x +_\star y$.

Answer 5.8.

$$\left(1, 8, 2, 4, \frac{1}{2}\right).$$

Problem 5.2. Let

$$\lambda = 3,$$

$$x = (2, 4, 6, 7).$$

Answer 5.9.

$$\left(e^{\log 2 \log 3}, e^{2 \log 2 \log 3}, e^{\log 3 \log 6}, e^{\log 3 \log 7}\right).$$

Problem 5.3. Let

$$x = (3, 4),$$

$$y = (2, 3),$$

$$z = (2, 5).$$

Find

$$(2 \cdot_\star x +_\star y) -_\star 3 \cdot_\star z.$$

Answer 5.10.

$$\left(2, e^{2(\log 2)^2 - \log 3 \log 5}\right).$$

Problem 5.4. Prove that

$$e_1 = \left(e^{\frac{1}{2}}, e^{\frac{1}{4}}, e^{\frac{1}{8}}\right)$$

$$e_2 = \left(e, e^{\frac{1}{2}}, e^{\frac{1}{4}}\right)$$

$$e_3 = \left(e^2, e, e^{\frac{1}{2}}\right)$$

are multiplicative linearly dependent and linearly independent.

Problem 5.5. Prove that

$$e_1 = \left(\frac{1}{3}, \frac{1}{6}, \frac{1}{9}\right)$$

$$e_2 = \left(1, \frac{1}{2}, \frac{1}{3}\right)$$

$$e_3 = \left(3, \frac{3}{2}, 1\right)$$

are multiplicative linearly independent and linearly dependent.

Problem 5.6. Let

$$e_1 = (4,5)$$

$$e_2 = (1,3).$$

Prove that e_1 and e_2 are multiplicative linearly independent and linearly independent.

Problem 5.7. Let

$$x = \left(e^{-1}, e^{-2}, e^3\right),$$

$$y = \left(e^{-2}, e^{-1}, e^2\right).$$

Find $\langle x, y \rangle_\star$.

Answer 5.11. e^{10}.

Problem 5.8. Let

$$x = \left(e^2, e^3, e, e^{-1}\right),$$

$$y = \left(e, e^2, e^2, e^4\right).$$

Check if $x \perp_\star y$.

Answer 5.12. *No.*

Problem 5.9. Let

$$x = \left(e^{-2}, e^2, e^{-2}, e^2\right).$$

Find $|x|_\star$.

Answer 5.13. e^4.

Problem 5.10. Let

$$x = (2,4,3),$$

$$y = (1,2,3).$$

Find $|x -_\star y|_\star$.

Answer 5.14.

$$e^{\sqrt{2}\log 2}.$$

6

Partial Multiplicative Differentiation

In this chapter, we define partial multiplicative derivatives of first and higher order and investigate some of their properties. In the chapter, we are considered multiplicative differentials and they are given their expressions. We define multiplicative directional derivatives and deduct some of their properties. In this chapter, we have given some necessary conditions for existence of local extremum of a function.

6.1 Definition for Multiplicative Functions of Several Variables

Definition 6.1. Let $f : \mathbb{R}_\star^n \to \mathbb{R}$. We define

$$f_\star(x) = e^{f(\log x_1, \log x_2, \ldots, \log x_n)}, \quad x \in \mathbb{R}_\star^n.$$

Example 6.1. Let $n = 3$ and

$$f(x) = x_1 \cdot_\star x_2 \cdot_\star x_3, \quad x \in \mathbb{R}_\star^3.$$

Then

$$f(x) = e^{\log x_1 \log x_2 \log x_3}, \quad x \in \mathbb{R}_\star^3,$$

and

$$f_\star(x) = e^{e^{\log(\log x_1)\log(\log x_2)\log(\log x_3)}}, \quad x \in \mathbb{R}_\star^3.$$

Example 6.2. Let $n = 2$ and

$$f(x) = x_1 +_\star x_2, \quad x \in \mathbb{R}_\star^2.$$

We have

$$f(x) = x_1 x_2, \quad x \in \mathbb{R}_\star^3,$$

and

$$f_\star(x) = e^{\log x_1 \log x_2}, \quad x \in \mathbb{R}_\star^2.$$

Example 6.3. Let $n = 2$ and

$$f(x) = \sin x_1 + \cos x_2, \quad x \in \mathbb{R}_\star^2.$$

Then

$$f_\star(x) = e^{\sin(\log x_1) + \cos(\log x_2)}, \quad x \in \mathbb{R}_\star^2.$$

DOI: 10.1201/9781003299080-6

Exercise 6.1. Let $n = 3$ and

$$f(x) = x_1 + x_2 - x_3, \quad x \in \mathbb{R}_\star^3.$$

Find $f_\star(x), x \in \mathbb{R}_\star^3$.

Answer 6.1.

$$e^{\log x_1 + \log x_2 - \log x_3}, \quad x \in \mathbb{R}_\star^3.$$

6.2 Definition for Multiplicative Partial Derivatives

Suppose that $A \subseteq \mathbb{R}_\star^n$ and $f \in \mathscr{C}^1(A)$.

Definition 6.2. For $j \in \{1, \ldots, n\}$, we define

$$f_{x_j}^\star(x) = e^{\frac{x_j f_{x_j}(x)}{f(x)}}, \quad x \in \mathbb{R}_\star^n.$$

The set of all functions $f : A \to \mathbb{R}$ that are continuous on A and have continuous partial multiplicative derivatives on A is denoted by $\mathscr{C}_\star^1(A)$.

Example 6.4. Let

$$f(x) = x_1^2 + x_2^2, \quad x \in \mathbb{R}_\star^2.$$

We have

$$f_{x_1}(x) \quad = \quad 2x_1,$$

$$f_{x_2}(x) \quad = \quad 2x_2, \quad x \in \mathbb{R}_\star^2,$$

and

$$f_{x_1}^\star(x) \quad = \quad e^{\frac{x_1 f_{x_1}(x)}{f(x)}}$$

$$= \quad e^{\frac{2x_1^2}{x_1^2 + x_2^2}},$$

$$f_{x_2}^\star(x) \quad = \quad e^{\frac{x_2 f_{x_2}(x)}{f(x)}}$$

$$= \quad e^{\frac{2x_2^2}{x_1^2 + x_2^2}}, \quad x \in \mathbb{R}_\star^2.$$

Example 6.5. Let

$$f(x) = 2 + \sin(x_1^2) + 2x_2^2, \quad x \in \mathbb{R}_*^2.$$

We have

$$f_{x_1}(x) \quad = \quad 2x_1 \cos(x_1^2),$$

$$f_{x_2}(x) \quad = \quad 4x_2, \quad x \in \mathbb{R}_*^2,$$

and

$$f_{x_1}^\star(x) \quad = \quad e^{\frac{x_1 f_{x_1}(x)}{f(x)}}$$

$$= \quad e^{\frac{2x_1^2 \cos(x_1^2)}{2 + \sin(x_1^2) + 2x_2^2}},$$

$$f_{x_2}^\star(x) \quad = \quad e^{\frac{x_2 f_{x_2}(x)}{f(x)}}$$

$$= \quad e^{\frac{4x_2^2}{2 + \sin(x_1^2) + 2x_2^2}}, \quad x \in \mathbb{R}_*^2.$$

Example 6.6. Let

$$f(x_1, x_2) = e^{x_1} + (\log x_2)^2 + 3 - \sin(x_1 x_2), \quad x \in \mathbb{R}_*^2.$$

We have

$$f_{x_1}(x_1, x_2) \quad = \quad e^{x_1} - x_2 \cos(x_1 x_2),$$

$$f_{x_2}(x_1, x_2) \quad = \quad \frac{2}{x_2} \log x_2 - x_1 \cos(x_1 x_2), \quad x \in \mathbb{R}_*^2.$$

Hence,

$$f_{x_1}^\star(x_1, x_2) \quad = \quad e^{\frac{x_1(e^{x_1} - x_2 \cos(x_1 x_2))}{e^{x_1} + (\log x_2)^2 + 3 - \sin(x_1 x_2)}},$$

$$f_{x_2}^\star(x_1, x_2) \quad = \quad e^{\frac{2 \log x_2 - x_1 x_2 \cos(x_1 x_2)}{e^{x_1} + (\log x_2)^2 + 3 - \sin(x_1 x_2)}}, \quad (x_1, x_2) \in \mathbb{R}_*^2.$$

Exercise 6.2. Let

$$f(x_1, x_2) = e^{x_1 + x_2} + 3 + \sin(x_1 + x_2), \quad (x_1, x_2) \in \mathbb{R}_*^2.$$

Find $f_{x_1}^\star$ and $f_{x_2}^\star$.

Answer 6.2.

$$f_{x_1}^\star(x_1,x_2) \;=\; e^{\dfrac{x_1\left(e^{x_1+x_2}+\cos(x_1+x_2)\right)}{e^{x_1+x_2}+3+\sin(x_1+x_2)}},$$

$$f_{x_2}^\star(x_1,x_2) \;=\; e^{\dfrac{x_2\left(e^{x_1+x_2}+\cos(x_1+x_2)\right)}{e^{x_1+x_2}+3+\sin(x_1+x_2)}}, \quad (x_1,x_2)\in\mathbb{R}_\star^2.$$

Definition 6.3. Suppose that $f \in \mathscr{C}^2(\mathbb{R}^n)$. We define

$$f_{x_i x_j}^{\star\star} = \left(f_{x_i}^\star\right)_{x_j}^\star, \quad i,j \in \{1,\ldots,n\}.$$

The set of all functions that are continuous on A and have continuous partial multiplicative derivatives up to second order is denoted by $\mathscr{C}_\star^2(A)$.

As above, it is defined the set $\mathscr{C}_\star^k(A)$, $k \in \mathbb{N}$. We have

$$f_{x_i x_i}^{\star\star} \;=\; e^{x_i\left(\frac{f_{x_i}}{f}+x_i\frac{f_{x_i x_i}f - f_{x_i}^2}{f^2}\right)},$$

$$f_{x_i x_j}^{\star\star} \;=\; e^{x_i x_j \frac{f_{x_i x_j}f - f_{x_i} f_{x_j}}{f^2}}, \quad i,j \in \{1,\ldots,n\}. \quad i \ne j.$$

Example 6.7. Let

$$f(x_1,x_2) = x_1^2 + x_2^2 + x_1 x_2, \quad (x_1,x_2) \in \mathbb{R}_\star^2.$$

We have

$$f_{x_1}(x_1,x_2) \;=\; 2x_1 + x_2,$$

$$f_{x_2}(x_1,x_2) \;=\; 2x_2 + x_1,$$

$$f_{x_1 x_1}(x_1,x_2) \;=\; 2,$$

$$f_{x_1 x_2}(x_1,x_2) \;=\; 1,$$

$$f_{x_2 x_2}(x_1,x_2) \;=\; 2, \quad (x_1,x_2) \in \mathbb{R}_\star^2.$$

Hence,

$$f_{x_1}^\star(x_1,x_2) \;=\; e^{\frac{x_1(2x_1+x_2)}{x_1^2+x_2^2+x_1 x_2}},$$

$$f_{x_2}^\star(x_1,x_2) \;=\; e^{\frac{x_2(2x_2+x_1)}{x_1^2+x_2^2+x_1 x_2}}, \quad (x_1,x_2) \in \mathbb{R}_\star^2,$$

and

$$f_{x_1x_1}^{\star\star}(x_1,x_2) \;=\; e^{x_1\left(\frac{2x_1+x_2}{x_1^2+x_2^2+x_1x_2}+x_1\frac{2x_1^2+2x_2^2+2x_1x_2-(2x_1+x_2)^2}{(x_1^2+x_2^2+x_1x_2)^2}\right)}$$

$$=\; e^{x_1\left(\frac{2x_1+x_2}{x_1^2+x_2^2+x_1x_2}+x_1\frac{-2x_1^2+x_2^2-2x_1x_2}{(x_1^2+x_2^2+x_1x_2)^2}\right)}$$

$$=\; e^{x_1\left(\frac{2x_1^3+2x_1x_2^2+2x_1^2x_2+x_2x_1^2+x_2^3+x_1x_2^2-2x_1^3+x_1x_2^2-2x_1^2x_2}{(x_1^2+x_2^2+x_1x_2)^2}\right)}$$

$$=\; e^{x_1\left(\frac{4x_1x_2^2+x_1^2x_2+x_2^3}{(x_1^2+x_2^2+x_1x_2)^2}\right)}, \quad (x_1,x_2)\in\mathbb{R}_\star^2,$$

and

$$f_{x_1x_2}^{\star\star}(x_1,x_2) \;=\; e^{x_1x_2\left(\frac{x_1^2+x_2^2+x_1x_2-(2x_1+x_2)(2x_2+x_1)}{(x_1^2+x_2^2+x_1x_2)^2}\right)}$$

$$=\; e^{x_1x_2\left(\frac{x_1^2+x_2^2+x_1x_2-4x_1x_2-2x_1^2-2x_2^2-x_1x_2}{(x_1^2+x_2^2+x_1x_2)^2}\right)}$$

$$=\; e^{-x_1x_2\left(\frac{x_1^2+4x_1x_2+x_2^2}{(x_1^2+x_2^2+x_1x_2)^2}\right)}, \quad (x_1,x_2)\in\mathbb{R}_\star^2,$$

and

$$f_{x_2x_2}^{\star\star}(x_1,x_2) \;=\; e^{x_2\left(\frac{2x_2+x_1}{x_1^2+x_2^2+x_1x_2}+x_2\frac{2x_1^2+2x_2^2+2x_1x_2-(2x_2+x_1)^2}{(x_1^2+x_2^2+x_1x_2)^2}\right)}$$

$$=\; e^{x_2\left(\frac{x_1+2x_2}{x_1^2+x_2^2+x_1x_2}+x_2\frac{2x_1^2+2x_2^2+2x_1x_2-4x_2^2-4x_1x_2-x_1^2}{(x_1^2+x_2^2+x_1x_2)^2}\right)}$$

$$=\; e^{x_2\left(\frac{2x_2+x_1}{x_1^2+x_2^2+x_1x_2}+x_2\frac{x_1^2-2x_2^2-2x_1x_2}{(x_1^2+x_2^2+x_1x_2)^2}\right)}$$

$$=\; e^{x_2\left(\frac{2x_1^2x_2+2x_2^3+2x_1x_2^2+x_1^3+x_1x_2^2+x_1^2x_2+x_1^2x_2-2x_2^3-2x_1x_2^2}{(x_1^2+x_2^2+x_1x_2)^2}\right)}$$

$$=\; e^{x_2\left(\frac{x_1^3+x_1x_2^2+4x_1^2x_2}{(x_1^2+x_2^2+x_1x_2)^2}\right)}, \quad (x_1,x_2)\in\mathbb{R}_\star^2.$$

Example 6.8. Let $A = \left[\dfrac{\pi}{4},\dfrac{\pi}{3}\right]\times\left[\dfrac{\pi}{4},\dfrac{\pi}{3}\right]$. Let also,

$$f(x_1,x_2) = \sin(x_1x_2), \quad (x_1,x_2)\in A.$$

We have

$$f_{x_1}(x_1, x_2) \;=\; x_2 \cos(x_1 x_2),$$

$$f_{x_2}(x_1, x_2) \;=\; x_1 \cos(x_1 x_2),$$

$$f_{x_1 x_1}(x_1, x_2) \;=\; -x_2^2 \sin(x_1 x_2),$$

$$f_{x_1 x_2}(x_1, x_2) \;=\; \cos(x_1 x_2) - x_1 x_2 \sin(x_1 x_2),$$

$$f_{x_2 x_2}(x_1, x_2) \;=\; -x_1^2 \sin(x_1 x_2), \quad (x_1, x_2) \in A.$$

Hence,

$$f_{x_1}^\star(x_1, x_2) \;=\; e^{x_1 x_2 \frac{\cos(x_1 x_2)}{\sin(x_1 x_2)}}$$

$$\;=\; e^{x_1 x_2 \cot(x_1 x_2)},$$

$$f_{x_2}^\star(x_1, x_2) \;=\; e^{x_1 x_2 \frac{\cos(x_1 x_2)}{\sin(x_1 x_2)}}$$

$$\;=\; e^{x_1 x_2 \cot(x_1 x_2)}, \quad (x_1, x_2) \in A,$$

and

$$f_{x_1 x_1}^{\star\star}(x_1, x_2) \;=\; e^{x_1 \left(x_2 \frac{\cos(x_1 x_2)}{\sin(x_1 x_2)} + x_1 \frac{-x_2^2 (\sin(x_1 x_2))^2 - x_2^2 (\cos(x_1 x_2))^2}{(\sin(x_1 x_2))^2} \right)}$$

$$\;=\; e^{x_1 \left(\frac{x_2 \cos(x_1 x_2)}{\sin(x_1 x_2)} - \frac{x_1 x_2^2}{(\sin(x_1 x_2))^2} \right)}$$

$$\;=\; e^{x_1 \left(\frac{x_2 \cos(x_1 x_2) \sin(x_1 x_2) - x_1 x_2^2}{(\sin(x_1 x_2))^2} \right)}$$

$$\;=\; e^{x_1 x_2 \left(\frac{\frac{1}{2} \sin(2 x_1 x_2) - x_1 x_2}{(\sin(x_1 x_2))^2} \right)}, \quad x_1, x_2) \in A,$$

and

$$f_{x_1 x_2}^{\star\star}(x_1, x_2) \;=\; e^{x_1 x_2 \left(\frac{\sin(x_1 x_2) \cos(x_1 x_2) - x_1 x_2 (\sin(x_1 x_2))^2 - x_1 x_2 (\cos(x_1 x_2))^2}{(\sin(x_1 x_2))^2} \right)}$$

$$\;=\; e^{x_1 x_2 \left(\frac{\frac{1}{2} \sin(x_1 x_2) - x_1 x_2}{(\sin(x_1 x_2))^2} \right)}, \quad (x_1, x_2) \in A,$$

and

$$f_{x_2x_2}^{\star\star}(x_1,x_2) = e^{x_2\left(\frac{x_1\cos(x_1x_2)}{\sin(x_1x_2)}+x_2\frac{-x_1^2(\sin(x_1x_2))^2-x_1^2(\cos(x_1x_2))^2}{(\sin(x_1x_2))^2}\right)}$$

$$= e^{x_2\left(\frac{x_1\sin(x_1x_2)\cos(x_1x_2)-x_1^2x_2}{(\sin(x_1x_2))^2}\right)}$$

$$= e^{x_1x_2\left(\frac{\frac{1}{2}\sin(2x_1x_2)-x_1x_2}{(\sin(x_1x_2))^2}\right)}, \quad (x_1,x_2) \in A.$$

Example 6.9. Let

$$f(x_1,x_2) = \log(x_1+x_2), \quad (x_1,x_2) \in \mathbb{R}_\star^2.$$

We have

$$f_{x_1}(x_1,x_2) = f_{x_2}(x_1,x_2)$$

$$= \frac{1}{x_1+x_2},$$

$$f_{x_1x_1}(x_1,x_2) = f_{x_1x_2}(x_1,x_2)$$

$$= f_{x_2x_2}(x_1,x_2)$$

$$= -\frac{1}{(x_1+x_2)^2}, \quad (x_1,x_2) \in \mathbb{R}_\star^2.$$

Hence,

$$f_{x_1}^{\star}(x_1,x_2) = e^{\frac{x_1}{(x_1+x_2)\log(x_1+x_2)}},$$

$$f_{x_2}^{\star}(x_1,x_2) = e^{\frac{x_2}{(x_1+x_2)\log(x_1+x_2)}}, \quad (x_1,x_2) \in \mathbb{R}_\star^2,$$

and

$$f_{x_1x_1}^{\star\star}(x_1,x_2) = e^{x_1\left(\frac{1}{(x_1+x_2)\log(x_1+x_2)}-x_1\frac{1+\log(x_1+x_2)}{(x_1+x_2)^2(\log(x_1+x_2))^2}\right)}$$

$$= e^{x_1\left(\frac{x_1\log(x_1+x_2)+x_2\log(x_1+x_2)-x_1-x_1\log(x_1+x_2)}{(x_1+x_2)^2(\log(x_1+x_2))^2}\right)}$$

$$= e^{x_1\frac{x_2\log(x_1+x_2)-x_1}{(x_1+x_2)^2(\log(x_1+x_2))^2}}, \quad (x_1,x_2) \in \mathbb{R}_\star^2,$$

and

$$f_{x_1 x_2}^{\star\star} x_1, x_2) \;=\; e^{-x_1 x_2 \frac{1+\log(x_1+x_2)}{(x_1+x_2)^2 (\log(x_1+x_2))^2}}, \quad (x_1, x_2) \in \mathbb{R}_\star^2,$$

and

$$f_{x_2 x_2}^{\star\star}(x_1, x_2) \;=\; e^{x_2 \left(\frac{1}{(x_1+x_2)\log(x_1+x_2)} - x_2 \frac{1+\log(x_1+x_2)}{(x_1+x_2)^2 (\log(x_1+x_2))^2} \right)}$$

$$\;=\; e^{x_2 \left(\frac{(x_1+x_2)\log(x_1+x_2) - x_2 - x_2\log(x_1+x_2)}{(x_1+x_2)^2 (\log(x_1+x_2))^2} \right)}$$

$$\;=\; e^{x_2 \frac{x_1 \log(x_1+x_2) - x_2}{(x_1+x_2)^2 (\log(x_1+x_2))^2}}, \quad (x_1, x_2) \in \mathbb{R}_\star^2.$$

Exercise 6.3. Let

$$f(x_1, x_2) = x_1 + x_2 + \sin(x_1 + x_2), \quad (x_1, x_2) \in \mathbb{R}_\star^2.$$

Find

$$f_{x_1}^{\star}(x_1, x_2), \quad f_{x_1 x_2}^{\star\star}(x_1, x_2), \quad (x_1, x_2) \in \mathbb{R}_\star^2.$$

Answer 6.3.

$$f_{x_1}^{\star}(x_1, x_2) \;=\; e^{\frac{x_1(1+\cos(x_1+x_2))}{x_1+x_2+\sin(x_1+x_2)}},$$

$$f_{x_1 x_2}^{\star\star}(x_1, x_2) \;=\; e^{x_1 x_2 \left(\frac{-(x_1+x_2)\sin(x_1+x_2) - 2\cos(x_1+x_2) - 2}{(x_1+x_2+\sin(x_1+x_2))^2} \right)}, \quad (x_1, x_2) \in \mathbb{R}_\star^2.$$

Note that, for any $f \in \mathscr{C}_\star^2(\mathbb{R}_\star^n)$, we have

$$f_{x_i x_j}^{\star\star}(x) = f_{x_j x_i}^{\star\star}(x), \quad x \in \mathbb{R}_\star^n,$$

$i, j \in \{1, \ldots, n\}$.

6.3 Multiplicative Differentials

Suppose that $A \subseteq \mathbb{R}_\star^n$ and f is enough times continuously-multiplicative differentiable on A.

Definition 6.4. We define

$$d_\star f(x) \;=\; f_{x_1}^{\star} \cdot_\star d_\star x_1 +_\star f_{x_2}^{\star} \cdot_\star d_\star x_2 +_\star \cdots +_\star f_{x_n}^{\star} \cdot_\star d_\star x_n$$

and

$$d_\star^2 f(x) = f_{x_1x_1}^{\star\star} {}^{\textstyle\cdot}_\star d_\star x_1^{2\star} +_\star f_{x_2x_2}^{\star\star} {}^{\textstyle\cdot}_\star d_\star x_2^{2\star} +_\star \cdots +_\star f_{x_nx_n}^{\star\star} {}^{\textstyle\cdot}_\star d_\star x_n^{2\star}$$

$$+_\star e^2 {}^{\textstyle\cdot}_\star f_{x_1x_2}^{\star\star} {}^{\textstyle\cdot}_\star d_\star x_1 {}^{\textstyle\cdot}_\star d_\star x_2 +_\star e^2 {}^{\textstyle\cdot}_\star f_{x_1x_3}^{\star\star} {}^{\textstyle\cdot}_\star d_\star x_1 {}^{\textstyle\cdot}_\star d_\star x_3$$

$$+_\star \cdots +_\star e^2 {}^{\textstyle\cdot}_\star f_{x_1x_n}^{\star\star} {}^{\textstyle\cdot}_\star d_\star x_1 {}^{\textstyle\cdot}_\star d_\star x_n$$

$$+_\star \cdots +_\star$$

$$+_\star e^2 {}^{\textstyle\cdot}_\star f_{x_{n-1}x_n}^{\star\star} {}^{\textstyle\cdot}_\star d_\star x_{n-1} {}^{\textstyle\cdot}_\star d_\star x_n,$$

and so on.

Example 6.10. Let

$$f(x_1,x_2) = x_1 x_2, \quad (x_1,x_2) \in \mathbb{R}_\star^2.$$

Then

$$f_{x_1}(x_1,x_2) = x_2,$$

$$f_{x_2}(x_1,x_2) = x_1,$$

$$f_{x_1x_1}(x_1,x_2) = 0,$$

$$f_{x_1x_2}(x_1,x_2) = 0,$$

$$f_{x_2x_2}(x_1,x_2) = 0, \quad (x_1,x_2) \in \mathbb{R}_\star^2.$$

Hence,

$$f_{x_1}^\star(x_1,x_2) = e^{x_1 \frac{x_2}{x_1x_2}}$$

$$= e,$$

$$f_{x_2}^\star(x_1,x_2) = e^{x_2 \frac{x_1}{x_1x_2}}$$

$$= e,$$

$$f_{x_1x_1}^\star(x_1,x_2) = e^{x_1\left(\frac{x_2}{x_1x_2} + x_1 \frac{0-x_2^2}{x_1^2x_2^2}\right)}$$

$$= e^{x_1\left(\frac{1}{x_1}-\frac{1}{x_1}\right)}$$

$$= e^0$$

$$= 1,$$

$$f^{\star\star}_{x_1x_2}(x_1,x_2) = e^{x_1x_2\left(\frac{x_1x_2-x_1x_2}{x_1^2x_2^2}\right)}$$

$$= e^0$$

$$= 1,$$

$$f^{\star\star}_{x_2x_2}(x_1,x_2) = e^{x_2\left(\frac{x_1}{x_1x_2}+x_2\frac{0-x_1^2}{x_1^2x_2^2}\right)}$$

$$= e^{x_2\left(\frac{1}{x_2}-\frac{1}{x_2}\right)}$$

$$= e^0$$

$$= 1, \quad (x_1,x_2) \in \mathbb{R}^2_\star.$$

Therefore,

$$d_\star f(x) = e \cdot_\star d_\star x_1 +_\star e \cdot_\star d_\star x_2,$$

$$d_\star^2 f(x) = 1 \cdot_\star d_\star x_1^{2\star} +_\star e^2 \cdot_\star 1 \cdot_\star d_\star x_1 \cdot_\star d_\star x_2 +_\star 1 \cdot_\star d_\star x_2^{2\star}$$

$$= 0_\star.$$

Example 6.11. Let

$$f(x_1,x_2) = \log(x_1 -_\star x_2), \quad (x_1,x_2) \in \mathbb{R}^2_\star.$$

Then

$$f(x_1,x_2) = \log\left(\frac{x_1}{x_2}\right), \quad (x_1,x_2) \in \mathbb{R}^2_\star,$$

and

$$f_{x_1}(x_1,x_2) = \frac{x_2}{x_1}\frac{1}{x_2}$$

$$= \frac{1}{x_1},$$

$$f_{x_2}(x_1, x_2) = \frac{x_2}{x_1}\left(-\frac{x_1}{x_2^2}\right)$$

$$= -\frac{1}{x_2},$$

$$f_{x_1 x_1}(x_1, x_2) = -\frac{1}{x_1^2},$$

$$f_{x_1 x_2}(x_1, x_2) = 0,$$

$$f_{x_2 x_2}(x_1, x_2) = \frac{1}{x_2^2}, \quad (x_1, x_2) \in \mathbb{R}_\star^2.$$

Hence,

$$f_{x_1}^\star(x_1, x_2) = e^{x_1 \frac{\frac{1}{x_1}}{\log\left(\frac{x_1}{x_2}\right)}}$$

$$= e^{\frac{1}{\log\left(\frac{x_1}{x_2}\right)}},$$

$$f_{x_2}^\star(x_1, x_2) = e^{x_2 \frac{-\frac{1}{x_2}}{\log\left(\frac{x_1}{x_2}\right)}}$$

$$= e^{-\frac{1}{\log\left(\frac{x_1}{x_2}\right)}}, \quad (x_1, x_2) \in \mathbb{R}_\star^2,$$

and

$$f_{x_1 x_1}^{\star\star}(x_1, x_2) = e^{x_1 \left(\frac{\frac{1}{x_1}}{\log\left(\frac{x_1}{x_2}\right)} + x_1 \frac{-\frac{1}{x_1^2}\log\left(\frac{x_1}{x_2}\right) - \frac{1}{x_1}}{\left(\log\left(\frac{x_1}{x_2}\right)\right)^2} \right)}$$

$$= e^{x_1 \left(\frac{1}{x_1 \log\left(\frac{x_1}{x_2}\right)} - \frac{\log\left(\frac{x_1}{x_2}\right) + 1}{x_1 \left(\log\left(\frac{x_1}{x_2}\right)\right)^2} \right)}$$

$$= e^{x_1 \frac{\log\left(\frac{x_1}{x_2}\right) - \log\left(\frac{x_1}{x_2}\right) - 1}{x_1 \left(\log\left(\frac{x_1}{x_2}\right)\right)^2}}$$

$$= e^{-\frac{1}{\left(\log\left(\frac{x_1}{x_2}\right)\right)^2}}, \quad (x_1, x_2) \in \mathbb{R}_\star^2,$$

and

$$f_{x_1x_2}^{\star\star}(x_1,x_2) = e^{x_1x_2\left(\frac{\frac{1}{x_1x_2}}{\left(\log\left(\frac{x_1}{x_2}\right)\right)^2}\right)}$$

$$= e^{\frac{1}{\left(\log\left(\frac{x_1}{x_2}\right)\right)^2}}, \quad x_1,x_2) \in \mathbb{R}_\star^2,$$

and

$$f_{x_2x_2}^{\star\star}(x_1,x_2) = e^{x_2\left(-\frac{1}{x_2\left(\log\left(\frac{x_1}{x_2}\right)\right)^2}\right)}$$

$$= e^{-\frac{1}{\left(\log\left(\frac{x_1}{x_2}\right)\right)^2}}, \quad (x_1,x_2) \in \mathbb{R}_\star^2.$$

Therefore,

$$d_\star f(x) = e^{\frac{1}{\log\left(\frac{x_1}{x_2}\right)}} \cdot_\star d_\star x_1 +_\star e^{-\frac{1}{\log\left(\frac{x_1}{x_2}\right)}} \cdot_\star d_\star x_2,$$

$(x_1,x_2) \in \mathbb{R}_\star^2$, and

$$d_\star^2 f(x) = e^{-\frac{1}{\left(\log\left(\frac{x_1}{x_2}\right)\right)^2}} \cdot_\star d_\star x_1^{2\star} +_\star e^2 \cdot_\star e^{\frac{1}{\left(\log\left(\frac{x_1}{x_2}\right)\right)^2}} \cdot_\star d_\star x_1 \cdot_\star d_\star x_2$$

$$+_\star e^{-\frac{1}{\left(\log\left(\frac{x_1}{x_2}\right)\right)^2}} \cdot_\star d_\star x_2^{2\star}, \quad (x_1,x_2) \in \mathbb{R}_\star^2.$$

Example 6.12. Let

$$f(x_1,x_2) = (x_1 + 2x_2)^2, \quad (x_1,x_2) \in \mathbb{R}_\star^2.$$

Then

$$f_{x_1}(x_1,x_2) = 2(x_1 + 2x_2),$$

$$f_{x_2}(x_1,x_2) = 4(x_1 + 2x_2),$$

$$f_{x_1x_1}(x_1,x_2) = 2,$$

$$f_{x_1x_2}(x_1,x_2) = 4,$$

$$f_{x_2x_2}(x_1,x_2) = 8, \quad (x_1,x_2) \in \mathbb{R}_\star^2.$$

Hence,

$$f_{x_1}^\star(x_1,x_2) = e^{x_1\frac{2(x_1+2x_2)}{(x_1+2x_2)^2}}$$

$$= e^{\frac{2x_1}{x_1+2x_2}},$$

$$f^{\star}_{x_2}(x_1,x_2) = e^{x_2\frac{4(x_1+2x_2)}{(x_1+2x_2)^2}}$$

$$= e^{\frac{4x_2}{x_1+2x_2}}, \quad (x_1,x_2) \in \mathbb{R}^2_{\star},$$

and

$$f^{\star\star}_{x_1x_1}(x_1,x_2) = e^{x_1\left(\frac{2}{x_1+2x_2} - \frac{2x_1}{(x_1+2x_2)^2}\right)}$$

$$= e^{\frac{4x_1x_2}{(x_1+2x_2)^2}}, \quad (x_1,x_2) \in \mathbb{R}^2_{\star},$$

and

$$f^{\star\star}_{x_1x_2}(x_1,x_2) = e^{-\frac{4x_1x_2}{(x_1+2x_2)^2}}, \quad (x_1,x_2) \in \mathbb{R}^2_{\star},$$

and

$$f^{\star\star}_{x_2x_2}(x_1,x_2) = e^{\frac{4x_1x_2}{(x_1+2x_2)^2}}, \quad (x_1,x_2) \in \mathbb{R}^2_{\star}.$$

Therefore,

$$d_{\star}f(x) = e^{\frac{2x_1}{x_1+2x_2}} \cdot_{\star} d_{\star}x +_{\star} e^{\frac{4x_2}{x_1+2x_2}} \cdot_{\star} d_{\star}x_2, \quad (x_1,x_2) \in \mathbb{R}^2_{\star},$$

and

$$d_{\star}^2 f(x) = e^{\frac{4x_1x_2}{(x_1+2x_2)^2}} \cdot_{\star} d_{\star}x_1^{2\star} +_{\star} e^2 \cdot_{\star} e^{-\frac{4x_1x_2}{(x_1+2x_2)^2}} \cdot_{\star} d_{\star}x_1 \cdot_{\star} d_{\star}x_2$$

$$+_{\star} e^{\frac{4x_1x_2}{(x_1+2x_2)^2}} \cdot_{\star} d_{\star}x_2^{2\star}, \quad (x_1,x_2) \in \mathbb{R}^2_{\star}.$$

Exercise 6.4. Let
$$f(x) = e^{x_1+x_2}, \quad (x_1,x_2) \in \mathbb{R}^2_{\star}.$$

Find
$$d_{\star}f(x_1,x_2), \quad d_{\star}^2 f(x_1,x_2), \quad (x_1,x_2) \in \mathbb{R}^2_{\star}.$$

Answer 6.4.

$$d_{\star}f(x_1,x_2) = e^{x_1} \cdot_{\star} d_{\star}x_1 +_{\star} e^{x_2} \cdot_{\star} d_{\star}x_2,$$

$$d_{\star}^2 f(x_1,x_2) = e^{x_1} \cdot_{\star} d_{\star}x_1^{2\star} +_{\star} e^{x_2} \cdot_{\star} d_{\star}x_2^{2\star}, \quad (x_1,x_2) \in \mathbb{R}^2_{\star}.$$

Below, suppose that f and g are enough times multiplicative continuously-differentiable on A. Let $a, b \in \mathbb{R}_\star$. Then

1. $d_\star(a \cdot_\star f(x) \pm_\star b \cdot_\star g(x)) = a \cdot_\star d_\star f(x) \pm_\star b \cdot_\star d_\star g(x), x \in A.$

2. $d_\star(f \cdot_\star g)(x) = f(x) \cdot_\star d_\star g(x) +_\star d_\star f(x) \cdot_\star g(x), x \in A.$

3. $d_\star(f/_\star g)(x) = (d_\star f(x) \cdot_\star g(x) -_\star f(x) \cdot_\star d_\star g(x))/_\star (g(x))^{2\star}, \; g(x) \neq 0_\star,$
   $x \in A.$

6.4   The Chain Rule

Suppose that $A \subseteq \mathbb{R}^2_\star$, $\alpha, \beta \in \mathbb{R}_\star$, $\alpha < \beta$. Let $g_j \in \mathscr{C}^1_\star([\alpha, \beta])$, $j \in \{1, \ldots, n\}$,

$$g(t) = (g_1(t), g_2(t), \ldots, g_n(t)) : [\alpha, \beta] \to A.$$

Let also, $f \in \mathscr{C}^1_\star(A)$. Then

$$d_\star(f(g(t)))/_\star d_\star t = e^{\frac{t}{f(g(t))} \frac{d}{dt} f(g(t))}$$

$$= e^{\frac{t}{f(g(t))} \left(f_{x_1}(g(t))g'_1(t) + f_{x_2}(g(t))g'_2(t) + \cdots + f_{x_n}(g(t))g'_n(t) \right)}$$

$$= e^{t \frac{f_{x_1}(g(t))}{f(g(t))} g'_1(t) + t \frac{f_{x_2}(g(t))}{f(g(t))} g'_2(t) + \cdots + t \frac{f_{x_n}(g(t))}{f(g(t))} g'_n(t)}$$

$$= \left(e^{t \frac{f_{x_1}(g(t))}{f(g(t))}} \right)^{g'_1(t)} \cdot \left(e^{t \frac{f_{x_2}(g(t))}{f(g(t))}} \right)^{g'_2(t)} \cdots \left(e^{t \frac{f_{x_n}(g(t))}{f(g(t))}} \right)^{g'_n(t)}$$

$$= \left(e^{g_1(t) \frac{f_{x_1}(g(t))}{f(g(t))} t \frac{g'_1(t)}{g_1(t)}} \right) \cdot \left(e^{g_2(t) \frac{f_{x_2}(g(t))}{f(g(t))} t \frac{g'_2(t)}{g_2(t)}} \right) \cdots \left(e^{g_n(t) \frac{f_{x_n}(g(t))}{f(g(t))} t \frac{g'_n(t)}{g_n(t)}} \right)$$

$$= \left(f^\star_{x_1}(g(t)) \right) \cdot_\star g^\star_1(t) +_\star \left(f^\star_{x_2}(g(t)) \right) \cdot_\star g^\star_2(t) +_\star \cdots$$
$$+_\star \left(f_{x_n}(g(t)) \right) \cdot_\star g^\star_n(t), \quad t \in [\alpha, \beta].$$

Example 6.13. Let

$$f(t) = (t^2 + t)^3 + (t^3 + t^2)^{10}, \quad t \in \mathbb{R}_\star.$$

Set

$$x_1(t) \quad = \quad t^2 + t,$$

$$x_2(t) = t^3 + t^2,$$

$$x(t) = (x_1(t), x_2(t)), \quad t \in \mathbb{R}_\star.$$

We have

$$x_1'(t) = 2t + 1,$$

$$x_2'(t) = 3t^2 + 2t, \quad t \in \mathbb{R}_\star,$$

$$f_{x_1}(x_1, x_2) = 3x_1^2,$$

$$f_{x_2}(x_1, x_2) = 10x_2^9, \quad (x_1, x_2) \in \mathbb{R}_\star^2.$$

Hence,

$$f_{x_1}^\star(x_1(t), x_2(t)) = e^{x_1(t) \frac{3(x_1(t))^2}{(x_1(t))^3 + (x_2(t))^{10}}}$$

$$= e^{\frac{3(t^2+t)^3}{(t^2+t)^3 + (t^3+t^2)^{10}}},$$

$$f_{x_2}^\star(x_1(t), x_2(t)) = e^{x_2(t) \frac{10(x_2(t))^9}{(x_1(t))^3 + (x_2(t))^{10}}}$$

$$= e^{\frac{10(t^3+t^2)^{10}}{(t^2+t)^3 + (t^3+t^2)^{10}}}$$

$$x_1^\star(t) = e^{\frac{t(2t+1)}{t^2+t}}$$

$$= e^{\frac{2t+1}{t+1}},$$

$$x_2^\star(t) = e^{t \frac{3t^2+2t}{t^3+t^2}}$$

$$= e^{\frac{3t+2}{t+1}}, \quad t \in \mathbb{R}_\star.$$

Therefore,

$$d_\star f(t)/_\star d_\star t = \left(f_{x_1}^\star(x_1(t), x_2(t))\right) \cdot_\star x_1^\star(t) +_\star \left(f_{x_2}^\star(x_1(t), x_2(t))\right) \cdot_\star x_2^\star(t)$$

$$= e^{\frac{3(t^2+1)^5}{(t^2+t)^3 + (t^3+t^2)^{10}}} \cdot_\star e^{\frac{2t+1}{t+1}} +_\star e^{\frac{10(t^3+t^2)^{10}}{(t^2+t)^3 + (t^3+t^2)^{10}}} \cdot_\star e^{\frac{3t+2}{t+1}}$$

$$= e^{\frac{3(t^2+1)^5(2t+1)}{(t+1)((t^2+t)^3+(t^3+t^2)^{10})} + \frac{10(t^3+t^2)^{10}(3t+2)}{(t+1)\left((t^2+t)^3+(t^3+t^2)^{10}\right)}}, \quad t \in \mathbb{R}_\star.$$

Example 6.14. Let

$$f(t) = (\sin t)^2(\cos t)^3, \quad t \in \left[\frac{\pi}{4},\frac{\pi}{3}\right].$$

Set

$$x_1(t) = \sin t,$$

$$x_2(t) = \cos t, \quad t \in \left[\frac{\pi}{4},\frac{\pi}{3}\right].$$

Then

$$x_1'(t) = \cos t,$$

$$x_2'(t) = -\sin t, \quad t \in \left[\frac{\pi}{4},\frac{\pi}{3}\right],$$

and

$$f(x_1(t),x_2(t)) = (x_1(t))^2(x_2(t))^3, \quad t \in \left[\frac{\pi}{4},\frac{\pi}{3}\right].$$

We have

$$f_{x_1}(x_1(t),x_2(t)) = 2x_1(t)(x_2(t))^3,$$

$$f_{x_2}(x_1(t),x_2(t)) = 3(x_1(t))^2(x_2(t))^2,$$

$$f_{x_1}^\star(x_1(t),x_2(t)) = e^2,$$

$$f_{x_2}^\star(x_1(t),x_2(t)) = e^3,$$

$$x_1^\star(t) = e^{t\cot t},$$

$$x_2^\star(t) = e^{-t\tan t}, \quad t \in \left[\frac{\pi}{4},\frac{\pi}{2}\right].$$

Hence,

$$d_\star f(x_1(t),x_2(t))/_\star d_\star t = \left(f_{x_1}^\star(x_1(t),x_2(t))\right)\cdot_\star x_1^\star(t) +_\star \left(f_{x_2}^\star(x_1(t),x_2(t))\right)\cdot_\star x_2^\star(t)$$

$$= e^2\cdot_\star e^{t\cot t} +_\star e^3\cdot_\star e^{-t\tan t}$$

$$= e^{2t\cot t - 3t\tan t}, \quad t \in \left[\frac{\pi}{4},\frac{\pi}{3}\right].$$

Example 6.15. Let

$$f(x_1(t), x_2(t)) = \left(e^t + e^{3t}\right)^3 t^4, \quad t \in \mathbb{R}_\star.$$

Set

$$x_1(t) = e^t + e^{3t},$$

$$x_2(t) = t, \quad t \in \mathbb{R}_\star.$$

Then

$$x_1'(t) = e^t + 3e^{3t},$$

$$x_2'(t) = 1, \quad t \in \mathbb{R}_\star,$$

and

$$f(x_1(t), x_2(t)) = (x_1(t))^3 (x_2(t))^4, \quad t \in \mathbb{R}_\star.$$

Hence,

$$f_{x_1}(x_1(t), x_2(t)) = 3(x_1(t))^2 (x_2(t))^3,$$

$$f_{x_2}(x_1(t), x_2(t)) = 4(x_1(t))^3 (x_2(t))^3, \quad t \in \mathbb{R}_\star,$$

and

$$f_{x_1}^\star(x_1(t), x_2(t)) = e^3,$$

$$f_{x_2}^\star(x_1(t), x_2(t)) = e^4,$$

$$x_1^\star(t) = e^{t \frac{e^t + 3e^{3t}}{e^t + e^{3t}}},$$

$$x_2^\star(t) = e \quad t \in \mathbb{R}_\star,$$

and

$$d_\star(f(t))/_\star d_\star t = \left(f_{x_1}^\star(x_1(t), x_2(t))\right) \cdot_\star x_1^\star(t) +_\star \left(f_{x_2}^\star(x_1(t), x_2(t))\right) \cdot_\star x_2^\star(t)$$

$$= e^3 \cdot_\star e^{t \frac{e^t + 3e^{3t}}{e^t + e^{3t}}} +_\star e^4 \cdot_\star e$$

$$= e^{3t \frac{e^t + 3e^{3t}}{e^t + e^{3t}} + 4}, \quad t \in \mathbb{R}_\star.$$

Exercise 6.5. Let

$$f(t) = (e^t + 2e^{-t})^4 (\cos t)^2, \quad t \in \mathbb{R}_\star.$$

Find

$$d_\star(f(t))/_\star d_\star t, \quad t \in \mathbb{R}_\star.$$

Answer 6.5.

$$e^{4t \frac{e^t - 2e^{-t}}{e^t + 2e^{-t}} - 2t \tan t}, \quad t \in \mathbb{R}_\star.$$

6.5   Multiplicative Homogeneous Functions

Let $A \subseteq \mathbb{R}_\star^n$.

Definition 6.5. Let $f : A \to \mathbb{R}_\star$. We say that f is a multiplicative homogeneous function if

$$f(t \cdot_\star x_1, t \cdot_\star x_2, \ldots, t \cdot_\star x_n) = t^{k_\star} \cdot_\star f(x_1, x_2, \ldots, x_n), \quad (x_1, x_2, \ldots, x_n) \in \mathbb{R}_\star^n. \quad (6.1)$$

$t \in \mathbb{R}_\star$, for some $k \in \mathbb{R}$. The number k is called the degree of multiplicative homogeneity.

We can rewrite equation (6.1) in the following way.

$$f\left(e^{\log t \log x_1}, e^{\log t \log x_2}, \ldots, e^{\log t \log x_n}\right) = e^{(\log t)^k \log f(x_1, x_2, \ldots, x_n)},$$

$(x_1, x_2, \ldots, x_n) \in A, t \in \mathbb{R}_\star.$

Example 6.16. Let

$$f(x_1, x_2) = x_1 x_2, \quad (x_1, x_2) \in \mathbb{R}_\star^2.$$

We have

$$
\begin{aligned}
f\left(e^{\log t \log x_1}, e^{\log t \log x_2}\right) &= e^{\log t \log x_1} e^{\log t \log x_2} \\[2mm]
&= e^{\log t \log x_1 + \log t \log x_2} \\[2mm]
&= e^{\log t (\log x_1 + \log x_2)} \\[2mm]
&= e^{\log t \log(x_1 x_2)} \\[2mm]
&= e^{\log t \log f(x_1, x_2)}, \quad (x_1, x_2) \in \mathbb{R}_\star^2, \quad t \in \mathbb{R}_\star.
\end{aligned}
$$

Thus, f is a multiplicative homogeneous function with degree of multiplicative homogeneity 1. Also, it is a homogeneous function with degree of homogeneity 2, because

$$f(tx_1, tx_2) = (tx_1)(tx_2)$$

$$= t^2 x_1 x_2$$

$$= t^2 f(x_1, x_2), \quad (x_1, x_2) \in \mathbb{R}_\star^2, \quad t \in \mathbb{R}_\star.$$

Example 6.17. Let

$$f(x_1, x_2) = x_1 + x_2, \quad (x_1, x_2) \in \mathbb{R}_\star^2.$$

We have

$$f(tx_1, tx_2) = tx_1 + tx_2$$

$$= t(x_1 + x_2)$$

$$= tf(x_1, x_2), \quad (x_1, x_2) \in \mathbb{R}_\star^2, \quad t \in \mathbb{R}_\star,$$

i.e., f is a homogeneous function with degree of homogeneity 1. Next,

$$f\left(e^{\log t \log x_1}, e^{\log t \log x_2}\right) = e^{\log t \log x_1} + e^{\log t \log x_2},$$

$(x_1, x_2) \in \mathbb{R}_\star^2, t \in \mathbb{R}_\star$. Hence, f is not a multiplicative homogeneous function.

Exercise 6.6. Let

$$f(x_1, x_2) = x_1^2 x_2^3, \quad (x_1, x_2) \in \mathbb{R}_\star^2.$$

Prove that f is a multiplicative homogeneous function.

Theorem 6.1 (The Euler Equality). *Let $f \in \mathscr{C}_\star^1(\mathbb{R}\star^n)$ be a multiplicative homogeneous function with degree of multiplicative homogeneity $k \in \mathbb{R}$. Then*

$$x_1 \cdot_\star \left(f_{x_1}^\star(x_1, x_2, \ldots, x_n)\right) +_\star x_2 \cdot_\star \left(f_{x_2}^\star(x_1, x_2, \ldots, x_n)\right) +_\star \cdots +_\star x_n \cdot_\star \left(f_{x_n}^\star(x_1, x_2, \ldots, x_n)\right)$$
$$= e^k \cdot_\star f(x_1, x_2, \ldots, x_n), \quad (x_1, x_2, \ldots, x_n) \in \mathbb{R}_\star^n.$$

Proof We multiplicative differentiate the equation (6.1) witn respect to t and we find

$$x_1 \cdot_\star \left(f_{x_1}^\star \left(e^{\log t \log x_1}, e^{\log t \log x_1}, \ldots, e^{\log t \log x_n}\right)\right)$$

$$+_\star x_2 \cdot_\star \left(f^\star_{x_2} \left(e^{\log t \log x_1}, e^{\log t \log x_2}, \ldots, e^{\log t \log x_n} \right) \right)$$

$$+_\star \cdots$$

$$+_\star x_n \cdot_\star \left(f^\star_{x_n} \left(e^{\log t \log x_1}, e^{\log t \log x_2}, \ldots, e^{\log t \log x_n} \right) \right)$$

$$= \quad e^{k \log f(x_1, \ldots, x_n)(\log t)^{k-1}}, \quad t \in \mathbb{R}_\star.$$

We put $t = e$ in the last equality and we get the desired equality. This completes the proof.

6.6   Multiplicative Directional Derivatives

Let $A \subseteq \mathbb{R}^n_\star$, $f : A \to \mathbb{R}_\star$, $f \in \mathscr{C}^1_\star(A)$. Let also, $s \in \mathbb{R}^n_\star$, $s = (s_1, s_2, \ldots, s_n)$.

Definition 6.6. Let $x \in A$. The multiplicative directional derivative of f at x in the multiplicative direction s we define as follows

$$\partial^\star f /_\star \partial^\star s(x) = \lim_{t \to 0_\star} \left(f(x +_\star t \cdot_\star s) -_\star f(x) \right) /_\star t. \tag{6.2}$$

Note that

$$
\begin{aligned}
x +_\star t \cdot_\star s &= (x_1, x_2, \ldots, x_n) +_\star t \cdot_\star (s_1, s_2, \ldots, s_n) \\
&= (x_1, x_2, \ldots, x_n) +_\star (t \cdot_\star s_1, t \cdot_\star s_2, \ldots, t \cdot_\star s_n) \\
&= (x_1 +_\star t \cdot_\star s_1, x_2 +_\star t \cdot_\star s_2, \ldots, x_n +_\star t \cdot_\star s_n) \\
&= \left(x_1 +_\star e^{\log t \log s_1}, x_2 +_\star e^{\log t \log s_2}, \ldots, x_n +_\star e^{\log t \log s_n} \right) \\
&= \left(x_1 e^{\log t \log s_1}, x_2 e^{\log t \log s_2}, \ldots, x_n e^{\log t \log s_n} \right)
\end{aligned}
$$

and

$$f(x +_\star t \cdot_\star s) = f \left(x_1 e^{\log t \log s_1}, x_2 e^{\log t \log s_2}, \ldots, x_n e^{\log t \log s_n} \right),$$

$$f(x +_\star t \cdot_\star s) -_\star f(x) = \frac{f \left(x_1 e^{\log t \log s_1}, x_2 e^{\log t \log s_2}, \ldots, x_n e^{\log t \log s_n} \right)}{f(x_1, x_2, \ldots, x_n)}$$

and

$$(f(x +_\star t \cdot_\star s) -_\star f(x)) /_\star t$$

$$= e^{\frac{\log f\left(x_1 e^{\log t \log s_1}, x_2 e^{\log t \log s_2}, \ldots, x_n e^{\log t \log s_n} \right) - \log f(x_1, x_2, \ldots, x_n)}{\log t}}.$$

Hence,

$$\lim_{t \to 0_\star} \left(f(x +_\star t \cdot_\star s) -_\star f(x) \right) /_\star t = \lim_{t \to 1} \left(f(x +_\star t \cdot_\star s) -_\star f(x) \right) /_\star t$$

$$= \quad f_{x_1}^\star(x) \cdot_\star s_1 +_\star f_{x_2}^\star(x) \cdot_\star s_2 +_\star \cdots +_\star f_{x_n}^\star(x) \cdot_\star s_n.$$

Therefore,

$$\partial_\star f /_\star \partial_\star s(x)$$
$$= \quad f_{x_1}^\star(x) \cdot_\star s_1 +_\star f_{x_2}^\star(x) \cdot_\star s_2 +_\star \cdots +_\star f_{x_n}^\star(x) \cdot_\star s_n.$$

Example 6.18. Let

$$f(x_1, x_2) \quad = \quad x_1^2 + x_2, \quad (x_1, x_2) \in \mathbb{R}_\star^2,$$

$$s \quad = \quad (2, 4).$$

Then

$$f_{x_1}(x_1, x_2) \quad = \quad 2x_1,$$

$$f_{x_2}(x_1, x_2) \quad = \quad 1, \quad (x_1, x_2) \in \mathbb{R}_\star^2.$$

Hence,

$$f_{x_1}^\star(x) \quad = \quad e^{x_1 \frac{2x_1}{x_1^2 + x_2}}$$

$$= \quad e^{\frac{2x_1^2}{x_1^2 + x_2}},$$

$$f_{x_2}^\star(x) \quad = \quad e^{x_2 \frac{1}{x_1^2 + x_2}}$$

$$= \quad e^{\frac{x_2}{x_1^2 + x_2}}, \quad (x_1, x_2) \in \mathbb{R}_\star^2.$$

Therefore,

$$(\partial_\star f /_\star \partial_\star s)(x) \quad = \quad \left(e^{\frac{2x_1^2}{x_1^2 + x_2}} \right) \cdot_\star 2 +_\star \left(e^{\frac{x_2}{x_1^2 + x_2}} \right) \cdot_\star 4$$

$$= \quad e^{\frac{2 \log 2x_1^2}{x_1^2 + x_2}} +_\star e^{\frac{2 \log 2x_2}{x_1^2 + x_2}}$$

$$= e^{\frac{2\log 2(x_1^2+x_2)}{x_1^2+x_2}}$$

$$= 4.$$

Example 6.19. Let

$$f(x_1,x_2) = 3+\sin x_1 + \cos x_2, \quad (x_1,x_2) \in \mathbb{R}_\star^2,$$

$$s = (4,8).$$

We have

$$f_{x_1}(x_1,x_2) = \cos x_1,$$

$$f_{x_2}(x_1,x_2) = -\sin x_2, \quad (x_1,x_2) \in \mathbb{R}_\star^2.$$

Hence,

$$f_{x_1}^\star(x_1,x_2) = e^{\frac{x_1 \cos x_1}{3+\sin x_1+\cos x_2}},$$

$$f_{x_2}^\star(x_1,x_2) = e^{-\frac{x_2 \sin x_2}{3+\sin x_1+\cos x_2}}, \quad (x_1,x_2) \in \mathbb{R}_\star^2,$$

and

$$
\begin{aligned}
(\partial_\star f/_\star \partial_\star s) &= \left(f_{x_1}^\star(x_1,x_2)\right) \cdot_\star s_1 +_\star \left(f_{x_2}^\star(x_1,x_2)\right) \cdot_\star s_2 \\
&= \left(e^{\frac{x_1 \cos x_1}{3+\sin x_1+\cos x_2}}\right) \cdot_\star 4 +_\star \left(e^{-\frac{x_2 \sin x_2}{3+\sin x_1+\cos x_2}}\right) \cdot_\star 8 \\
&= e^{\frac{2\log 2 x_1 \cos x_1}{3+\sin x_1+\cos x_2}} +_\star e^{-\frac{3\log 2 x_2 \sin x_2 \log 2}{3+\sin x_1+\cos x_2}} \\
&= e^{\frac{2\log 2(x_1 \cos x_1 - \frac{3}{2}x_2 \sin x_2)}{3+\sin x_1+\cos x_2}}, \quad (x_1,x_2) \in \mathbb{R}_\star^2.
\end{aligned}
$$

Example 6.20. Let

$$f(x_1,x_2) = \log(x_1 x_2), \quad (x_1,x_2) \in \mathbb{R}_\star^2,$$

$$s = (3,4).$$

Then

$$f_{x_1}(x_1,x_2) = \frac{x_2}{x_1 x_2}$$

$$= \frac{1}{x_1},$$

$$f_{x_2}(x_1,x_2) = \frac{x_1}{x_1 x_2}$$

$$= \frac{1}{x_2}, \quad (x_1,x_2) \in \mathbb{R}^2_\star,$$

and

$$f^\star_{x_1}(x_1,x_2) = e,$$

$$f^\star_{x_2}(x_1,x_2) = e, \quad (x_1,x_2) \in \mathbb{R}^2_\star,$$

and

$$(\partial_\star f/{}_\star \partial_\star s)(x) = \left(f^\star_{x_1}(x)\right) \cdot_\star s_1 +_\star \left(f^\star_{x_2}(x)\right) \cdot_\star s_2$$

$$= e \cdot_\star 3 +_\star e \cdot_\star 4$$

$$= 3 +_\star 4$$

$$= 12, \quad (x_1,x_2) \in \mathbb{R}^2_\star.$$

Exercise 6.7. Let

$$f(x_1,x_2) = x_1 x_2^2, \quad (x_1,x_2) \in \mathbb{R}^2_\star,$$

$$s = (2,5).$$

Find

$$(\partial_\star f/{}_\star \partial_\star s)(x).$$

Answer 6.6.

$$2e^{2\log 5}.$$

6.7 Extremum of a Function

Suppose that $A \subseteq \mathbb{R}^n_\star$.

Theorem 6.2. *Let* $f : A \to \mathbb{R}_\star$ *has a local extremum at* $x^0 \in A$. *Then* $f^\star_{x_j}(x^0)$, $j \in \{1,\ldots,n\}$, *do not exist or*

$$f^\star_{x_j}(x^0) = 0_\star, \quad j \in \{1,\ldots,n\}.$$

Proof Since f has a local extremum at x^0, we have that $f_{x_j}(x^0)$, $j \in \{1,\dots,n\}$, do not exist or

$$f_{x_j}(x^0) = 0, \quad j \in \{1,\dots,n\}.$$

Since

$$f^\star_{x_j}(x^0) = e^{x^0_j \frac{f_{x_j}(x^0)}{f(x^0)}}, \quad j \in \{1,\dots,n\},$$

we get that $f^\star_{x_j}(x^0)$, $j \in \{1,\dots,n\}$, do not exist or

$$f^\star_{x_j}(x^0) = 0_\star, \quad j \in \{1,\dots,n\}.$$

This completes the proof.

Corollary 6.1. If f is multiplicative continuously differentiable at x^0 and has a local extremum at x^0, then

$$f^\star_{x_j}(x^0) = 0_\star, \quad j \in \{1,\dots,n\}.$$

6.8   Advanced Practical Problems

Problem 6.1. Let $n = 3$ and

$$f(x) = \cos(x_1 x_2 x_3), \quad x \in \mathbb{R}^3_\star.$$

Find $f_\star(x)$, $x \in \mathbb{R}^3_\star$.

Answer 6.7.
$$f_\star(x) = \cos(\log x_1 \log x_2 \log x_3), \quad x \in \mathbb{R}^3_\star.$$

Problem 6.2. Let

$$f(x_1,x_2) = e^{x_1^2} + 3 + x_2^2 + \frac{1}{1+x_1^2+x_2^2}, \quad (x_1,x_2) \in \mathbb{R}^2_\star.$$

Find $f^\star_{x_1}$ and $f^\star_{x_2}$.

Answer 6.8.

$$f^\star_{x_1}(x_1,x_2) = e^{\frac{x_1\left(2x_1 e^{x_1^2} - \frac{2x_1}{(1+x_1^2+x_2^2)^2}\right)}{x_1^2+3+x_2^2+\frac{1}{1+x_1^2+x_2^2}}},$$

$$f^\star_{x_2}(x_1,x_2) = e^{\frac{x_2\left(2x_2 - \frac{2x_2}{(1+x_1^2+x_2^2)^2}\right)}{e^{x_1^2}+3+x_2^2+\frac{1}{1+x_1^2+x_2^2}}}, \quad (x_1,x_2) \in \mathbb{R}^2_\star.$$

Problem 6.3. Let
$$f(x_1, x_2) = x_1^2 x_2, \quad (x_1, x_2) \in \mathbb{R}_\star^2.$$

Find
$$f_{x_1}^\star(x_1, x_2), \quad f_{x_1 x_2}^{\star\star}(x_1, x_2), \quad (x_1, x_2) \in \mathbb{R}_\star^2.$$

Answer 6.9.

$$f_{x_1}^\star(x_1, x_2) = e^2,$$

$$f_{x_1 x_2}^{\star\star}(x_1, x_2) = 0_\star, \quad (x_1, x_2) \in \mathbb{R}_\star^2.$$

Problem 6.4. Let
$$f(x) = e^{x_1 - x_2}, \quad (x_1, x_2) \in \mathbb{R}_\star^2.$$

Find
$$d_\star f(x_1, x_2), \quad d_\star^2 f(x_1, x_2), \quad (x_1, x_2) \in \mathbb{R}_\star^2.$$

Answer 6.10.

$$d_\star f(x_1, x_2) = e^{x_1} \cdot_\star d_\star x_1 +_\star e^{-x_2} \cdot_\star d_\star x_2,$$

$$d_\star^2 f(x_1, x_2) = e^{x_1} \cdot_\star d_\star x_1^{2\star} +_\star e^{-x_2} \cdot_\star d_\star x_2^{2\star}, \quad (x_1, x_2) \in \mathbb{R}_\star^2.$$

Problem 6.5. Let
$$f(t) = \left(e^{2t} + e^{-3t}\right)^5 (\sin t)^3, \quad t \in \mathbb{R}_\star.$$

Find
$$d_\star(f(t))/_\star d_\star t, \quad t \in \mathbb{R}_\star.$$

Answer 6.11.

$$e^{5t \frac{2e^{2t} - 3e^{-3t}}{e^{2t} + e^{-3t}} + 3t \cot t}, \quad t \in \mathbb{R}_\star.$$

Problem 6.6. Let
$$f(x_1, x_2) = x_1^8 x_2^4, \quad (x_1, x_2) \in \mathbb{R}_\star^2.$$
Prove that f is a multiplicative homogeneous function.

Problem 6.7. Let
$$f(x_1, x_2) = x_1^7 x_2^6, \quad (x_1, x_2) \in \mathbb{R}_\star^2,$$

$$s = (9, 11).$$

Find
$$(\partial_\star f /_\star \partial_\star s)(x).$$

Answer 6.12.
$$e^{14 \log 3 + 6 \log 11}.$$

7

Multiple Multiplicative Integrals

In this chapter, we investigate multiplicative integrals depending on parameters and using them we define iterated multiplicative integrals. We define multiple multiplicative and multiple improper multiplicative integrals and we investigate their properties.

7.1 Multiplicative Integrals Depending on Parameters

Suppose that $a, b, a_j, b_j \in \mathbb{R}_\star$, $j \in \{1,\ldots,m\}$, $a < b$, $a_j < b_j$, $j \in \{1,\ldots,m\}$. Let

$$D = \{(x, \alpha_1, \alpha_2, \ldots, \alpha_m) : a \leq x \leq b, \quad a_j \leq \alpha_j \leq b_j, \quad j \in \{1,\ldots,m\}\}$$

and

$$D_1 = \{(\alpha_1, \alpha_2, \ldots, \alpha_m) : a_j \leq \alpha_j \leq b_j, \quad j \in \{1,\ldots,m\}\}.$$

Definition 7.1. Let $f : D \to \mathbb{R}_\star$, $f(\cdot, \alpha_1, \alpha_2, \ldots, \alpha_m) \in \mathscr{C}_\star([a,b])$ for any $(\alpha_1, \alpha_2, \ldots, \alpha_m) \in D_1$. Then there exists

$$F(\alpha_1, \alpha_2, \ldots, \alpha_m) = \int_{\star a}^{b} f(x, \alpha_1, \alpha_2, \ldots, \alpha_m) \cdot_\star d_\star x, \quad (\alpha_1, \alpha_2, \ldots, \alpha_m) \in D_1.$$

The function F is said to be a multiplicative integral depending on the parameters $\alpha_1, \alpha_2, \ldots, \alpha_m$.

Note that

$$F(\alpha_1, \alpha_2, \ldots, \alpha_m) = e^{\int_a^b \frac{1}{x} \log f(x, \alpha_1, \alpha_2, \ldots, \alpha_m) dx},$$

$(\alpha_1, \alpha_m, \ldots, \alpha_m) \in D_1$. If $f \in \mathscr{C}_\star(D)$, then $F \in \mathscr{C}_\star^2(D_1)$. Let now, $f(x, \cdot, \ldots, \cdot) \in \mathscr{C}_\star^1(D_1)$ for any $x \in [a,b]$. Then

$$F_{\alpha_j}^\star(\alpha_1, \alpha_2, \ldots, \alpha_m) = e^{\alpha_j \frac{F_{\alpha_j}(\alpha_1, \alpha_2, \ldots, \alpha_m)}{F(\alpha_1, \alpha_2, \ldots, \alpha_m)}}$$

$$= e^{\alpha_j \int_a^b \frac{1}{x} \frac{f_{\alpha_j}(x, \alpha_1, \alpha_2, \ldots, \alpha_m)}{f(x, \alpha_1, \alpha_2, \ldots, \alpha_m)} dx}$$

$$= e^{\int_a^b \frac{1}{x} \log f_{\alpha_j}^\star(x, \alpha_1, \alpha_2, \ldots, \alpha_m) dx}$$

$$= \int_{\star a}^{b} f_{\alpha_j}^\star(x, \alpha_1, \alpha_2, \ldots, \alpha_m) \cdot_\star d_\star x, \quad (\alpha_1, \alpha_2, \ldots, \alpha_m) \in D_1.$$

DOI: 10.1201/9781003299080-7

Example 7.1. Let
$$f(x, \alpha) = e^{x^2 + \alpha x}, \quad x, \alpha \in \mathbb{R}_\star.$$

Then

$$
\begin{aligned}
F(\alpha) &= \int_{\star 2}^4 f(x, \alpha) \cdot_\star d_\star x \\
&= e^{\int_2^4 \frac{1}{x} \log e^{x^2 + \alpha x} dx} \\
&= e^{\int_2^4 \frac{x^2 + \alpha x}{x} dx} \\
&= e^{\int_2^4 (x + \alpha) dx} \\
&= e^{\left. \left(\frac{x^2}{2} + \alpha x \right) \right|_{x=2}^{x=4}} \\
&= e^{8 + 4\alpha - 2 - 2\alpha} \\
&= e^{6 + 2\alpha}, \quad \alpha \in \mathbb{R}_\star.
\end{aligned}
$$

Hence,

$$
\begin{aligned}
F'(\alpha) &= 2e^{6 + 2\alpha}, \\
F^\star(\alpha) &= e^{\alpha \frac{F'(\alpha)}{F(\alpha)}} \\
&= e^{\alpha \frac{2e^{6+2\alpha}}{e^{6+2\alpha}}} \\
&= e^{2\alpha}, \quad \alpha \in \mathbb{R}_\star.
\end{aligned}
$$

Example 7.2. Let
$$f(x, \alpha) = e^{x \cos(x + \alpha)}, \quad x, \alpha \in \mathbb{R}_\star.$$

We have

$$
\begin{aligned}
F(\alpha) &= \int_{\star \frac{\pi}{4}}^{\frac{\pi}{2}} f(x, \alpha) \cdot_\star d_\star x \\
&= e^{\int_{\frac{\pi}{4}}^{\frac{\pi}{2}} \frac{1}{x} \log f(x, \alpha) dx}
\end{aligned}
$$

$$= e^{\int_{\frac{\pi}{4}}^{\frac{\pi}{2}} \frac{1}{x} x \cos(x+\alpha)dx}$$

$$= e^{\int_{\frac{\pi}{4}}^{\frac{\pi}{2}} \cos(x+\alpha)dx}$$

$$= e^{\sin(x+\alpha)\Big|_{x=\frac{\pi}{4}}^{x=\frac{\pi}{2}}}$$

$$= e^{\sin\left(\frac{\pi}{2}+\alpha\right)-\sin\left(\frac{\pi}{4}+\alpha\right)}$$

$$= e^{2\sin\frac{\pi}{8}\cos\left(\frac{3\pi}{8}+\alpha\right)}, \quad \alpha \in \mathbb{R}_\star.$$

Hence,

$$F'(\alpha) = -\sin\left(\frac{3\pi}{8}+\alpha\right)e^{2\sin\frac{\pi}{8}\cos\left(\frac{3\pi}{8}+\alpha\right)}, \quad \alpha \in \mathbb{R}_\star,$$

and

$$F^\star(\alpha) = e^{\alpha\frac{F'(\alpha)}{F(\alpha)}}$$

$$= e^{\alpha\frac{-\sin\left(\frac{3\pi}{8}+\alpha\right)e^{2\sin\frac{\pi}{8}\cos\left(\frac{3\pi}{8}+\alpha\right)}}{e^{2\sin\frac{\pi}{8}\cos\left(\frac{3\pi}{8}+\alpha\right)}}}$$

$$= e^{-\alpha\sin\left(\frac{3\pi}{8}+\alpha\right)}, \quad \alpha \in \mathbb{R}_\star.$$

Example 7.3. Let

$$f(x,\alpha_1,\alpha_2) = e^{x^2+\alpha_1^2 x+\alpha_2^3 x}, \quad x,\alpha_1,\alpha_2 \in \mathbb{R}_\star.$$

Then

$$F(\alpha_1,\alpha_2) = \int_{\star 1}^{2} f(x,\alpha_1,\alpha_2) \cdot_\star d_\star x$$

$$= e^{\int_1^2 \frac{1}{x} \log f(x,\alpha_1,\alpha_2)dx}$$

$$= e^{\int_1^2 \frac{1}{x}(x^2+\alpha_1^2 x+\alpha_2^3 x)dx}$$

$$= e^{\int_1^2 (x+\alpha_1^2+\alpha_2^3)dx}$$

$$= e^{\frac{1}{2}x^2\Big|_{x=1}^{x=2}+(\alpha_1^2+\alpha_2^3)x\Big|_{x=1}^{x=2}}$$

$$= e^{\frac{3}{2}+\alpha_1^2+\alpha_2^3}, \quad \alpha_1, \alpha_2 \in \mathbb{R}_\star.$$

Hence,

$$F_{\alpha_1}(\alpha_1, \alpha_2) = 2\alpha_1 e^{\frac{3}{2}+\alpha_1^2+\alpha_2^3},$$

$$F_{\alpha_2}(\alpha_1, \alpha_2) = 3\alpha_2^2 e^{\frac{3}{2}+\alpha_1^2+\alpha_2^3}, \quad \alpha_1, \alpha_2 \in \mathbb{R}_\star,$$

and

$$F_{\alpha_1}^\star(\alpha_1, \alpha_2) = e^{\alpha_1 \frac{F_{\alpha_1}(\alpha_1, \alpha_2)}{F(\alpha_1, \alpha_2)}}$$

$$= e^{2\alpha_1^2},$$

$$F_{\alpha_2}^\star(\alpha_1, \alpha_2) = e^{\alpha_2 \frac{F_{\alpha_2}(\alpha_1, \alpha_2)}{F(\alpha_1, \alpha_2)}}$$

$$= e^{3\alpha_2^3}, \quad \alpha_1, \alpha_2 \in \mathbb{R}_\star.$$

Exercise 7.1. Let $a = 1$, $b = 2$ and

$$f(x, \alpha_1, \alpha_2, \alpha_3) = e^{x^3 + (\alpha_1^2 + \alpha_2^3 + \alpha_3)x}, \quad x, \alpha_1, \alpha_2, \alpha_3 \in \mathbb{R}_\star.$$

Find

1. $F(\alpha_1, \alpha_2, \alpha_2)$, $\alpha_1, \alpha_2, \alpha_3 \in \mathbb{R}_\star$.
2. $F_{\alpha_1}^\star(\alpha_1, \alpha_2, \alpha_3)$, $\alpha_1, \alpha_2, \alpha_3 \in \mathbb{R}_\star$.
3. $F_{\alpha_2}^\star(\alpha_1, \alpha_2, \alpha_3)$, $\alpha_1, \alpha_2, \alpha_3 \in \mathbb{R}_\star$.
4. $F_{\alpha_3}^\star(\alpha_1, \alpha_2, \alpha_3)$, $\alpha_1, \alpha_2, \alpha_3 \in \mathbb{R}_\star$.

Answer 7.1.   *1.* $e^{\frac{7}{3}+\alpha_1^2+\alpha_2^3+\alpha_3}$, $\alpha_1, \alpha_2, \alpha_3 \in \mathbb{R}_\star$.

2. $e^{2\alpha_1^2}$, $\alpha_1, \alpha_2, \alpha_3 \in \mathbb{R}_\star$.

3. $e^{3\alpha_2^3}$, $\alpha_1, \alpha_2, \alpha_3 \in \mathbb{R}_\star$.

4. e^{α_3}, $\alpha_1, \alpha_2, \alpha_3 \in \mathbb{R}_\star$.

Below, suppose that $c, d \in \mathbb{R}_\star$, $c < d$.

Theorem 7.1. *Let $f \in \mathscr{C}_\star([a,b] \times [c,d])$, $f_\alpha^\star(x, \alpha)$ exists and it is continuous on* $[a, b] \times [c, d]$. *Let also,* $\phi, \psi : [c, d] \to [a, b]$, $\phi, \psi \in \mathscr{C}_\star^1([c, d])$, *and*

$$F(\alpha) = \int_{\star\phi(\alpha)}^{\psi(\alpha)} f(x, \alpha) \cdot_\star d_\star x.$$

Then

$$F^\star(\alpha) = \left((\psi^\star(\alpha))^{\log f(\psi(\alpha), \alpha)} -_\star (\phi^\star(\alpha))^{\log f(\phi(\alpha), \alpha)} \right) +_\star \int_{\star\phi(\alpha)}^{\psi(\alpha)} f_\alpha^\star(x, \alpha) \cdot_\star d_\star x.$$

Proof We have

$$F(\alpha) = e^{\int_{\phi(\alpha)}^{\psi(\alpha)} \frac{1}{x} \log f(x,\alpha) dx}, \quad \alpha \in [c,d],$$

and

$$F'(\alpha) = \left(\frac{\psi'(\alpha)}{\psi(\alpha)} \log f(\psi(\alpha), \alpha) - \frac{\phi'(\alpha)}{\phi(\alpha)} \log f(\phi(\alpha), \alpha) + \int_{\phi(\alpha)}^{\psi(\alpha)} \frac{1}{x} \frac{f_\alpha(x,\alpha)}{f(x,\alpha)} dx \right)$$

$$\cdot e^{\int_{\phi(\alpha)}^{\psi(\alpha)} \frac{1}{x} \log f(x,\alpha) dx}, \quad \alpha \in [c,d].$$

Hence,

$$F^\star(\alpha) = e^{\alpha \frac{F'(\alpha)}{F(\alpha)}}$$

$$= e^{\alpha \frac{\psi'(\alpha)}{\psi(\alpha)} \log f(\psi(\alpha),\alpha) - \alpha \frac{\phi'(\alpha)}{\phi(\alpha)} \log f(\phi(\alpha),\alpha) + \alpha \int_{\phi(\alpha)}^{\psi(\alpha)} \frac{1}{x} \frac{f_\alpha(x,\alpha)}{f(x,\alpha)} dx}$$

$$= e^{\alpha \frac{\psi'(\alpha)}{\psi(\alpha)} \log f(\psi(\alpha),\alpha) - \alpha \frac{\phi'(\alpha)}{\phi(\alpha)} \log f(\phi(\alpha),\alpha)} \cdot e^{\alpha \int_{\phi(\alpha)}^{\psi(\alpha)} \frac{1}{x} \frac{f_\alpha(x,\alpha)}{f(x,\alpha)} dx}$$

$$= \left((\psi^\star(\alpha))^{\log f(\psi(\alpha),\alpha)} -_\star (\phi^\star(\alpha))^{\log f(\phi(\alpha),\alpha)} \right) +_\star \int_{\star\phi(\alpha)}^{\psi(\alpha)} f_\alpha^\star(x,\alpha) \cdot_\star d_\star x,$$

$\alpha \in [c,d]$. This completes the proof.

7.2 Iterated Multiplicative Integrals

Let $a_j, b_j \in \mathbb{R}_\star$, $a_j < b_j$, $j \in \{1,\ldots,n\}$,

$$D = [a_1,b_1] \times [a_2,b_2] \times \cdots [a_n,b_n].$$

Definition 7.2. Let $f : D \to \mathbb{R}_\star$. Suppose that

$$f(\cdot, x_2, \ldots, x_n), \quad f(x_1, \cdot, x_3, \ldots, x_n), \quad \ldots, \quad f(x_1, x_2, \ldots, x_{n-1}, \cdot)$$

are multiplicative integrable on $[a_1,b_1]$, $[a_2,b_2]$, $\ldots$, $[a_n,b_n]$, respectively, for any $(x_1, x_2, \ldots, x_n) \in D$. We define

$$I = \int_{\star D} f(x_1, x_2, \ldots, x_n) \cdot_\star d_\star x_1 \cdot_\star \cdots d_\star x_{n-1} \cdot_\star d_\star x_n$$

$$= \int_{\star a_n}^{b_n} \int_{\star a_{n-1}}^{b_{n-1}} \cdots \int_{a_1}^{b_1} f(x_1, x_2, \ldots, x_n) \cdot_\star d_\star x_1 \cdot_\star \cdots \cdot_\star d_\star x_{n-1} \cdot_\star d_\star x_n$$

as follows

$$I = \int_{\star a_n}^{b_n} \left(\int_{\star a_{n-1}}^{b_{n-1}2} \cdots \left(\int_{\star a_2}^{b_2} \left(\int_{\star a_1}^{b_1} f(x_1, x_2, \ldots, x_n) \cdot_{\star} d_{\star} x_1 \right) \cdot_{\star} d_{\star} x_2 \right) \cdots \cdot_{\star} d_{\star} x_{n-1} \right) \cdot_{\star} d_{\star} x_n.$$

(7.1)

The multiplicative integral (7.1) will be called the iterated multiplicative integral.

We can rewrite the multiplicative integral (7.1) in the following manner

$$I = e^{\int_{a_n}^{b_n} \left(\int_{a_{n-1}}^{b_{n-1}} \cdots \left(\int_{a_2}^{b_2} \left(\int_{a_1}^{b_1} \frac{\log f(x_1, x_2, \ldots, x_n)}{x_1 x_2 \cdots x_n} dx_1 \right) dx_2 \right) \cdots dx_{n-1} \right) dx_n}.$$

Example 7.4. Let

$$f(x_1, x_2) = e^{x_1^2 x_2^2 + x_1 x_2}, \quad (x_1, x_2) \in [2, 3] \times [2, 3].$$

We will compute

$$I = \int_{\star 2}^{3} \int_{\star 2}^{3} f(x_1, x_2) \cdot_{\star} d_{\star} x_1 \cdot_{\star} d_{\star} x_2.$$

We have

$$
\begin{aligned}
I &= e^{\int_2^3 \left(\int_2^3 \frac{1}{x_1 x_2} \log f(x_1, x_2) dx_1 \right) dx_2} \\[2mm]
&= e^{\int_2^3 \left(\int_2^3 \frac{1}{x_1 x_2} \log e^{x_1^2 x_2^2 + x_1 x_2} dx_1 \right) dx_2} \\[2mm]
&= e^{\int_2^3 \left(\int_2^3 \frac{1}{x_1 x_2} (x_1^2 x_2^2 + x_1 x_2) dx_1 \right) dx_2} \\[2mm]
&= e^{\int_2^3 \left(\int_2^3 (x_1 x_2 + 1) dx_1 \right) dx_2} \\[2mm]
&= e^{\int_2^3 \left(x_2 \frac{x_1^2}{2} \Big|_{x_1=2}^{x_1=3} + x_1 \Big|_{x_1=2}^{x_1=3} \right) dx_2} \\[2mm]
&= e^{\int_2^3 \left(\frac{5}{2} x_2 + 1 \right) dx_2} \\[2mm]
&= e^{\frac{5}{4} x_2^2 \Big|_{x_2=2}^{x_2=3} + x_2 \Big|_{x_2=2}^{x_2=3}} \\[2mm]
&= e^{\frac{25}{4} + 1} \\[2mm]
&= e^{\frac{29}{4}}.
\end{aligned}
$$

Example 7.5. Let

$$f(x_1, x_2) = x_1 x_2, \quad (x_1, x_2) \in [1, 2] \times [1, 2].$$

We will compute

$$I = \int_{*1}^{2} \int_{*1}^{2} f(x_1, x_2) \cdot_\star d_\star x_1 \cdot_\star d_\star x_2.$$

We have

$$
\begin{aligned}
I &= e^{\int_1^2 \left(\int_1^2 \frac{1}{x_1 x_2} \log(x_1 x_2) dx_1 \right) dx_2} \\[2ex]
&= e^{\int_1^2 \frac{1}{x_2} \left(\int_1^2 \frac{1}{x_1} (\log x_1 + \log x_2) dx_1 \right) dx_2} \\[2ex]
&= e^{\int_1^2 \frac{1}{x_2} \left(\int_1^2 \frac{1}{x_1} \log x_1 dx_1 + \int_1^2 \frac{1}{x_1} \log x_2 dx_1 \right) dx_2} \\[2ex]
&= e^{\int_1^2 \frac{1}{x_2} \left(\int_1^2 \log x_1 d\log x_1 + \log x_2 \log x_1 \Big|_{x_1=1}^{x_1=2} \right) dx_2} \\[2ex]
&= e^{\int_1^2 \frac{1}{x_2} \left(\frac{(\log x_1)^2}{2} \Big|_{x_1=1}^{x_1=2} + \log x_2 (\log 2 - \log 1) \right) dx_2} \\[2ex]
&= e^{\int_1^2 \frac{1}{x_2} \left(\frac{(\log 2)^2}{2} + \log 2 \log x_2 \right) dx_2} \\[2ex]
&= e^{\frac{(\log 2)^2}{2} \int_1^2 \frac{1}{x_2} dx_2 + \log 2 \int_1^2 \frac{1}{x_2} \log x_2 dx_2} \\[2ex]
&= e^{\frac{(\log 2)^2}{2} \log x_2 \Big|_{x_2=1}^{x_2=2} + \log 2 \int_1^2 \log x_2 d\log x_2} \\[2ex]
&= e^{\frac{(\log 2)^3}{2} + \log 2 \frac{(\log x_2)^2}{2} \Big|_{x_2=1}^{x_2=2}} \\[2ex]
&= e^{\frac{(\log 2)^3}{2} + \frac{(\log 2)^3}{2}} \\[2ex]
&= e^{(\log 2)^3}.
\end{aligned}
$$

Example 7.6. Let

$$f(x_1, x_2, x_3) = e^{x_1 + x_2 + x_3}, \quad (x_1, x_2, x_3) \in [1, 2] \times [1, 2] \times [1, 2].$$

We will compute

$$I = \int_{\star 1}^{2} \left(\int_{\star 1}^{2} \left(\int_{\star 1}^{2} f(x_1, x_2, x_3) \cdot_\star d_\star x_1 \right) \cdot_\star d_\star x_2 \right) \cdot_\star d_\star x_3.$$

We have

$$I = e^{\int_1^2 \left(\int_1^2 \left(\int_1^2 \frac{1}{x_1 x_2 x_3} \log e^{x_1 + x_2 + x_3} dx_1 \right) dx_2 \right) dx_3}$$

$$= e^{\int_1^2 \left(\int_1^2 \left(\int_1^2 \frac{1}{x_1 x_2 x_3} (x_1 + x_2 + x_3) dx_1 \right) dx_2 \right) dx_3}$$

$$= e^{\int_1^2 \left(\int_1^2 \left(\int_1^2 \left(\frac{1}{x_2 x_3} + \frac{1}{x_1 x_3} + \frac{1}{x_1 x_2} \right) dx_1 \right) dx_2 \right) dx_3}$$

$$= e^{\int_1^2 \left(\int_1^2 \left(\frac{1}{x_2 x_3} x_1 \Big|_{x_1=1}^{x_1=2} + \frac{1}{x_3} \log x_1 \Big|_{x_1=1}^{x_1=2} + \frac{1}{x_2} \log x_1 \Big|_{x_1=1}^{x_1=2} \right) dx_2 \right) dx_3}$$

$$= e^{\int_1^2 \left(\int_1^2 \left(\frac{1}{x_2 x_3} + \frac{\log 2}{x_3} + \frac{\log 2}{x_2} \right) dx_2 \right) dx_3}$$

$$= e^{\int_1^2 \left(\frac{1}{x_3} \log x_2 \Big|_{x_2=1}^{x_2=2} + \frac{\log 2}{x_3} x_2 \Big|_{x_2=1}^{x_2=2} + \log 2 \log x_2 \Big|_{x_2=1}^{x_2=2} \right) dx_3}$$

$$= e^{\int_1^2 \left(\frac{\log 2}{x_3} + \frac{\log 2}{x_3} + (\log 2)^2 \right) dx_3}$$

$$= e^{\int_1^2 \left(\frac{2\log 2}{x_3} + (\log 2)^2 \right) dx_3}$$

$$= e^{2\log 2 \log x_3 \Big|_{x_3=1}^{x_3=2} + (\log 2)^2 x_3 \Big|_{x_3=1}^{x_3=2}}$$

$$= e^{2(\log 2)^2 + (\log 2)^2}$$

$$= e^{3(\log 2)^2}.$$

Exercise 7.2. Let

$$f(x_1, x_2, x_3) = e^{x_1 x_2 x_3}, \quad (x_1, x_2) \in [1, 2] \times [1, 2].$$

Find

$$I = \int_{\star 1}^{2} \left(\int_{\star 1}^{2} \left(\int_{\star 1}^{2} f(x_1, x_2, x_3) \cdot_\star d_\star x_1 \right) \cdot_\star d_\star x_2 \right) \cdot_\star d_\star x_3.$$

Answer 7.2. *e.*

Let $f, g : D \to \mathbb{R}_\star$, $a, b \in \mathbb{R}_\star$, and

$$f(\cdot, x_2, \ldots, x_n), \quad g(\cdot, x_2, \ldots, x_n); \quad f(x_1, \cdot, x_3, \ldots, x_n), \quad g(x_1, \cdot, x_3, \ldots, x_n);$$

$$\ldots$$

$$f(x_1, x_2, \ldots, x_{n-1}, \cdot), \quad g(x_1, x_2, \ldots, x_{n-1}, \cdot)$$

are multiplicative integrable functions on $[a_1, b_1]$, $[a_2, b_2]$, $\ldots$, $[a_n, b_n]$, respectively. Then

$$\int_{\star D} (a \cdot_\star f(x_1, x_2, \ldots, x_n) +_\star b \cdot_\star g(x_1, x_2, \ldots, x_n)) \cdot_\star d_\star x_1 \cdot_\star d_\star x_2 \cdot_\star \cdots_\star d_\star x_n$$

$$= \quad a \cdot_\star \int_{\star D} f(x_1, x_2, \ldots, x_n) \cdot_\star d_\star x_1 \cdot_\star d_\star x_2 \cdot_\star \cdots_\star d_\star x_n$$

$$+_\star b \cdot_\star \int_{\star D} g(x_1, x_2, \ldots, x_n) \cdot_\star d_\star x_1 \cdot_\star d_\star x_2 \cdot_\star \cdots_\star d_\star x_n.$$

7.3 Multiple Multiplicative Integrals

Let $D \subseteq \mathbb{R}_\star^n$.

Definition 7.3. Suppose that $f : D \to \mathbb{R}_\star$, $f \in \mathscr{C}_\star(D)$. We define multiple multiplicative integrals as follows

$$\int_{\star D} f(x) \cdot_\star d_\star x = e^{\int_D \frac{1}{x_1 x_2 \ldots x_n} f(x_1, x_2, \ldots, x_n) dx_1 dx_2 \ldots dx_n},$$

where

$$d_\star x = d_\star x_1 \cdot_\star d_\star x_2 \cdot_\star \ldots \cdot_\star d_\star x_n.$$

If $f, g : D \to \mathbb{R}_\star$, $f, g \in \mathscr{C}_\star(D)$, $a, b \in \mathbb{R}_\star$, then

$$\int_{\star D} (a \cdot_\star f(x) +_\star b \cdot_\star g(x)) \cdot_\star d_\star x = a \cdot_\star \int_{\star D} f(x) \cdot_\star d_\star x$$

$$+_\star b \cdot_\star \int_{\star D} g(x) \cdot_\star d_\star x.$$

Example 7.7. Let

$$D = \{(x_1, x_2) : x_2 \geq \frac{x_1}{2}, \quad x_2 \leq \frac{2}{x_1}, \quad x_1 \geq 1\}.$$

We will compute

$$I = \int_{\star D} e^{x_1^3 x_2^2} \cdot_\star d_\star x.$$

We have

$$
\begin{aligned}
I &= e^{\int \int_D \frac{1}{x_1 x_2} \log e^{x_1^3 x_2^2} dx_1 dx_2} \\[2mm]
&= e^{\int \int_D \frac{1}{x_1 x_2} (x_1^3 x_2^2) dx_1 dx_2} \\[2mm]
&= e^{\int \int_D x_1^2 x_2 dx_1 dx_2}.
\end{aligned}
$$

Let

$$J = \int \int_D x_1^2 x_2 dx_1 dx_2.$$

We have that

$$\frac{2}{x_1} \geq \frac{x_1}{2}, \quad x_1 \geq 1,$$

if and only if

$$4 \geq x_1^2, \quad x_1 \geq 1,$$

if and only if

$$2 \geq x_1 \geq 1.$$

Thus,

$$D = \{(x_1, x_2) \in \mathbb{R}_\star^2 : 1 \leq x_1 \leq 2, \quad \frac{x_1}{2} \leq x_2 \leq \frac{2}{x_1}\}.$$

Hence,

$$
\begin{aligned}
J &= \int_1^2 \int_{\frac{x_1}{2}}^{\frac{2}{x_1}} x_1^2 x_2 dx_2 dx_1 \\[2mm]
&= \int_1^2 x_1^2 \left(\int_{\frac{x_1}{2}}^{\frac{2}{x_1}} x_2 dx_2 \right) dx_1 \\[2mm]
&= \int_1^2 x_1^2 \frac{x_2^2}{2} \Big|_{x_2 = \frac{x_1}{2}}^{x_2 = \frac{2}{x_1}} dx_1 \\[2mm]
&= \int_1^2 x_1^2 \left(\frac{4}{2x_1^2} - \frac{x_1^2}{8} \right) dx_1 \\[2mm]
&= \int_1^2 \left(2 - \frac{x_1^4}{8} \right) dx_1 \\[2mm]
&= 2x_1 \Big|_{x_1=1}^{x_1=2} - \frac{x_1^5}{40} \Big|_{x_1=1}^{x_1=2}
\end{aligned}
$$

$$= 2 - \frac{32-1}{40}$$

$$= 2 - \frac{31}{40}$$

$$= \frac{49}{40}.$$

Therefore,

$$I = e^{\frac{49}{40}}.$$

Example 7.8. Let

$$D = \{(x_1, x_2) \in \mathbb{R}_\star^2 : x_1^2 + x_2^2 \leq 16\}.$$

We will compute

$$I = \int_{\star D} e^{x_1 x_2 (x_1^2 + x_2^2)} \cdot_\star d_\star x_1 \cdot_\star d_\star x_2.$$

We have

$$I = e^{\int\int_D \frac{1}{x_1 x_2} \log e^{x_1 x_2 (x_1^2 + x_2^2)} dx_1 dx_2}$$

$$= e^{\int\int_D (x_1^2 + x_2^2) dx_1 dx_2}.$$

Let

$$J = \int\int_D (x_1^2 + x_2^2) dx_1 dx_2.$$

Set

$$x_1 = r\cos\phi$$

$$x_2 = r\sin\phi, \quad r \geq 0, \quad \phi \in [0, 2\pi].$$

Since $(x_1, x_2) \in \mathbb{R}_\star^2$, then $\phi \in \left[0, \frac{\pi}{2}\right]$. Then

$$D = \{(x_1, x_2) \in \mathbb{R}_\star^2 : x_1 = r\cos\phi, \quad x_2 = r\sin\phi, \quad r \in [0,4], \quad \phi \in \left[0, \frac{\pi}{2}\right]\}.$$

Therefore,

$$J = \int_0^4 \int_0^{\frac{\pi}{2}} r^3 d\phi dr$$

$$= \frac{\pi}{2} \int_0^4 r^3 dr$$

$$= \frac{\pi}{2} \frac{r^4}{4} \Big|_{r=0}^{r=4}$$

$$= \frac{\pi}{8} \cdot 256$$

$$= 32\pi$$

and

$$I = e^{32\pi}.$$

Example 7.9. Let

$$D = \{(x_1, x_2, x_3) \in \mathbb{R}^3_\star : x_1^2 + x_2^2 + x_3^2 \leq 9\}.$$

We will compute

$$I = \int_{\star D} e^{x_1^2 x_2^2 x_3^2} \cdot_\star d_\star x_1 \cdot_\star d_\star x_2 \cdot_\star d_\star x_3.$$

We have

$$I = e^{\int \int \int_D \frac{1}{x_1 x_2 x_3} \log e^{x_1^2 x_2^2 x_3^2} dx_1 dx_2 dx_3}$$

$$= e^{\int \int \int_D x_1 x_2 x_3 dx_1 dx_2 dx_3}.$$

Let

$$J = \int \int \int_D x_1 x_2 x_3 dx_1 dx_2 dx_3.$$

Set

$$x_1 = r\cos\phi\sin\theta$$

$$x_2 = r\sin\phi\sin\theta$$

$$x_3 = r\cos\theta, \quad r \geq 0, \quad \phi \in [0, 2\pi], \quad \theta \in [0, \pi].$$

Since $(x_1, x_2, x_3) \in \mathbb{R}_\star$, we get that $\theta, \phi \in \left[0, \frac{\pi}{2}\right]$. Therefore,

$$J = \int_0^3 \int_0^{\frac{\pi}{2}} \int_0^{\frac{\pi}{2}} r^3 \cos\phi\sin\phi(\sin\theta)^2 \cos\theta r^2 \sin\theta d\theta d\phi dr$$

$$= \int_0^3 \int_0^{\frac{\pi}{2}} \int_0^{\frac{\pi}{2}} r^5 \cos\phi\sin\phi(\sin\theta)^3 \cos\theta d\theta d\phi dr$$

$$= \left(\int_0^3 r^5 dr\right)\left(\int_0^{\frac{\pi}{2}} \cos\phi\sin\phi d\phi\right)\left(\int_0^{\frac{\pi}{2}} (\sin\theta)^3 \cos\theta d\theta\right)$$

$$= \left(\frac{r^6}{6}\Big|_{r=0}^{r=3}\right)\left(\frac{(\sin\phi)^2}{2}\Big|_{\phi=0}^{\phi=\frac{\pi}{2}}\right)\left(\frac{(\sin\theta)^4}{4}\Big|_{\theta=0}^{\theta=\frac{\pi}{2}}\right)$$

$$= \frac{729}{6}\cdot\frac{1}{2}\cdot\frac{1}{4}$$

$$= \frac{243}{16}$$

and

$$I = e^{\frac{243}{16}}.$$

Exercise 7.3. Let

$$D = \{(x_1, x_2) \in \mathbb{R}_\star^2 : 1 \le x_1 \le 2, \quad 1 \le x_2 \le 2x_1\}.$$

Compute

$$\int_{\star D} e^{x_1^2 x_2^2} \cdot_\star d_\star x_1 \cdot_\star d_\star x_2.$$

7.4 Multiple Improper Multiplicative Integrals

Let D be a bounded measurable set in $\mathbb{R}_\star^n$.

Definition 7.4. Let $M_0 \in D$ and f be defined on $D\backslash M_0$, and $\lim\limits_{M\to M_0} f(M) = \infty$. Take $\varepsilon > 0$ so that $U(M_0, \varepsilon) \subset D$, and let $D_\varepsilon = D\backslash U(M_0, \varepsilon)$, where $U(M_0, \varepsilon)$ is a neighborhood of M_0. Assume that $f \in \mathscr{C}_\star(D_\varepsilon)$. If the limit

$$\lim_{\varepsilon\to 0}\int_{\star D_\varepsilon} f(x) \cdot_\star d_\star x = \int_{\star D} f(x) \cdot_\star d_\star x$$

exists, then it is said to be a multiple improper multiplicative integral. In this case, we say that it is convergent. Otherwise, we say that it is divergent.

Example 7.10. Let

$$D = \{(x_1, x_2) \in \mathbb{R}_\star^2 : 1 \le x_1^2 + x_2^2 \le 4\}.$$

Consider

$$I = \int_{\star D} e^{\frac{x_1 x_2}{(x_1^2 + x_2^2 - 1)^{\frac{1}{2}}}} \cdot_\star d_\star x_1 \cdot_\star d_\star x_2.$$

We have

$$D_\varepsilon = \{(x_1, x_2) \in \mathbb{R}_\star^2 : 1 + \varepsilon \le x_1^2 + x_2^2 \le 4\}.$$

Then

$$
\begin{aligned}
I &= \lim_{\varepsilon \to 0} e^{\int\int_{D_\varepsilon} \frac{1}{x_1 x_2} \log e^{\frac{x_1 x_2}{(x_1^2 + x_2^2 - 1)^2}} \, dx_1 \, dx_2} \\[2mm]
&= \lim_{\varepsilon \to 0} e^{\int\int_{D_\varepsilon} \frac{1}{(x_1^2 + x_2^2 - 1)^{\frac{1}{2}}} \, dx_1 \, dx_2} \\[2mm]
&= \lim_{\varepsilon \to 0} e^{\frac{\pi}{2} \int_{1+\varepsilon}^{2} \frac{\rho}{(\rho^2 - 1)^{\frac{1}{2}}} \, d\rho} \\[2mm]
&= \lim_{\varepsilon \to 0} e^{\frac{\pi}{2} (\rho^2 - 1)^{\frac{1}{2}} \Big|_{\rho = 1+\varepsilon}^{\rho = 2}} \\[2mm]
&= \lim_{\varepsilon \to 0} e^{\frac{\pi}{2} (1 - \varepsilon^{\frac{1}{2}})} \\[2mm]
&= e^{\frac{\pi}{2}}.
\end{aligned}
$$

Thus, the considered multiplicative integral is convergent.

Exercise 7.4. Let

$$
D = \{ (x_1, x_2) \in \mathbb{R}_*^2 : 1 \leq x_1^2 + x_2^2 \leq 4 \}.
$$

Prove that

$$
I = \int_{*D} e^{\frac{x_1 x_2}{(x_1^2 + x_2^2 - 1)^5}} \cdot_* d_* x_1 \cdot_* d_* x_2
$$

is divergent.

7.5   Advanced Practical Problems

Problem 7.1. Let $a = 1$, $b = 2$ and

$$
f(x, \alpha_1, \alpha_2, \alpha_3) = e^{x^4 + (\alpha_1^7 + \alpha_2^4 + \alpha_3^5)x}, \quad x, \alpha_1, \alpha_2, \alpha_3 \in \mathbb{R}_*.
$$

Find

1. $F(\alpha_1, \alpha_2, \alpha_2)$, $\alpha_1, \alpha_2, \alpha_3 \in \mathbb{R}_*$.
2. $F_{\alpha_1}^{\star}(\alpha_1, \alpha_2, \alpha_3)$, $\alpha_1, \alpha_2, \alpha_3 \in \mathbb{R}_*$.
3. $F_{\alpha_2}^{\star}(\alpha_1, \alpha_2, \alpha_3)$, $\alpha_1, \alpha_2, \alpha_3 \in \mathbb{R}_*$.
4. $F_{\alpha_3}^{\star}(\alpha_1, \alpha_2, \alpha_3)$, $\alpha_1, \alpha_2, \alpha_3 \in \mathbb{R}_*$.

Answer 7.3.   *1.* $e^{\frac{15}{4}+\alpha_1^7+\alpha_2^4+\alpha_3^5}$, $\alpha_1,\alpha_2,\alpha_3 \in \mathbb{R}_\star$.

2. $e^{7\alpha_1^7}$, $\alpha_1,\alpha_2,\alpha_3 \in \mathbb{R}_\star$.

3. $e^{4\alpha_2^4}$, $\alpha_1,\alpha_2,\alpha_3 \in \mathbb{R}_\star$.

4. $e^{5\alpha_3^5}$, $\alpha_1,\alpha_2,\alpha_3 \in \mathbb{R}_\star$.

Problem 7.2. Let

$$f(x_1,x_2,x_3) = e^{x_1^2 x_2^2 x_3^2}, \quad (x_1,x_2) \in [1,2] \times [1,2].$$

Find

$$I = \int_{\star 1}^{2} \left(\int_{\star 1}^{2} \left(\int_{\star 1}^{2} f(x_1,x_2,x_3) \cdot_\star d_\star x_1 \right) \cdot_\star d_\star x_2 \right) \cdot_\star d_\star x_3.$$

Answer 7.4. $e^{\frac{27}{8}}$.

Problem 7.3. Let

$$D = \{(x_1,x_2) \in \mathbb{R}_\star^2 : 1 \le x_1 \le 3, \quad 1 \le x_2 \le 3x_1^2\}.$$

Compute

$$\int_{\star D} e^{x_1^3 x_2^2} \cdot_\star d_\star x_1 \cdot_\star d_\star x_2.$$

Problem 7.4. Let

$$D = \{(x_1,x_2) \in \mathbb{R}_\star^2 : x_1^2 + x_2^2 \le 9\}.$$

Compute

$$\int_{\star D} e^{x_1^3 x_2 \sqrt{9-x_1^2-x_2^2}} \cdot_\star d_\star x_1 \cdot_\star d_\star x_2.$$

Problem 7.5. Let

$$D = \{(x_1,x_2) \in \mathbb{R}_\star^3 : 1 \le x_1 \le 2x_2, \quad 1 \le x_3 \le 4x_2^2\}.$$

Compute

$$\int_{\star D} e^{x_1^3 x_2^2 x_3^2} \cdot_\star d_\star x_1 \cdot_\star d_\star x_2 \cdot_\star x_3.$$

Problem 7.6. Let

$$D = \{(x_1,x_2) \in \mathbb{R}_\star^3 : 81 \le x_1^2 + x_2^2 + x_3^2 \le 256\}.$$

Compute

$$\int_{\star D} e^{x_1 x_2(x_1^2+x_2^2+x_3^2)} \cdot_\star d_\star x_1 \cdot_\star d_\star x_2 \cdot_\star d_\star x_3.$$

Problem 7.7. Let

$$D = \{(x_1,x_2) \in \mathbb{R}_\star^2 : 4 \le x_1^2 + x_2^2 \le 64\}.$$

Prove that

$$I = \int_{\star D} e^{\frac{x_1 x_2}{(x_1^2+x_2^2-4)^9}} \cdot_\star d_\star x_1 \cdot_\star d_\star x_2$$

is divergent.

References

[1] D. Aniszewska, Multiplicative Runge-Kutta Method, Nonlinear Dynamics 50 (1-2) (2007) 265-272.

[2] A. Bashirov, E. Kurpinar, A. Özyapici, Multiplicative Calculus and its Applications, Journal of Mathematical Analysis and its Applications 337 (1) (2008) 36-48.

[3] F. Córdova-Lepe, The Multiplicative Derivative as a Measure of Elasticity in Economics, TEMAT-Theaeteto Atheniensi Mathematica 2 (3) (2006), online.

[4] S. G. Georgiev, Focus on Calculus, Nova Science Publisher, 2020.

[5] B. Gompertz, On the Nature of the Function Expressive of the Law of Human Mortality, and on a New Mode of Determining the Value of Life Contingencies, Philosophical Transactions of the Royal Society of London 115 (1825) 513-585.

[6] M. Grossman, R. Katz, Non-Newtonian Calculus, Pigeon Cove, Lee Press, Massachusats, 1972.

[7] M. Grossman, Bigeometric Calculus: A System with a Scale-Free Derivative, Archimedes Foundation, Rockport, Massachusats, 1983.

[8] W. Kasprzak, B. Lysik, M. Rybaczuk, Dimensions, Invariants Models and Fractals, Ukrainian Society on Fracture Mechanics, SPOLOM, Wroclaw-Lviv, Poland, 2004.

[9] R.R. Meginniss, Non-Newtonian Calculus Applied to Probability, Utility, and Bayesian Analysis, Manuscript of the report for delivery at the 20th KDKR-KSF Seminar on Bayesian Inference in Econometrics, Purdue University, West Lafayette, Indiana, May 23, 1980.

[10] M. Riza, A. Özyapici, E. Misirli, Multiplicative Finite Difference Methods, Quarterly of Applied Mathematics, 67 (4) (2009) 745-754.

[11] M. Rybaczuk, A. Kedzia, W. Zielinski, The Concepts of Physical and Fractional Dimensions II. The Differential Calculus in Dimensional Spaces, Chaos Solutions Fractals 12 (2001) 2537-2552.

[12] D. Stanley, A Multiplicative Calculus, Primus IX (4) (1999) 310-326.

203

Index